ROUTLEDGE LIBRARY EDITIONS: ENVIRONMENTAL POLICY

Volume 4

THE GREENING OF MACHIAVELLI

THE GREENING OF MACHIAVELLI
The Evolution of International Environmental Politics

TONY BRENTON

LONDON AND NEW YORK

First published in 1994 by Earthscan Publications Ltd

This edition first published in 2019
by Routledge
2 Park Square, Milton Park, Abingdon, Oxon OX14 4RN

and by Routledge
52 Vanderbilt Avenue, New York, NY 10017

Routledge is an imprint of the Taylor & Francis Group, an informa business

© 1994 Royal Institute of International Affairs

All rights reserved. No part of this book may be reprinted or reproduced or utilised in any form or by any electronic, mechanical, or other means, now known or hereafter invented, including photocopying and recording, or in any information storage or retrieval system, without permission in writing from the publishers.

Trademark notice: Product or corporate names may be trademarks or registered trademarks, and are used only for identification and explanation without intent to infringe.

British Library Cataloguing in Publication Data
A catalogue record for this book is available from the British Library

ISBN: 978-0-367-18894-8 (Set)
ISBN: 978-0-429-27423-7 (Set) (ebk)
ISBN: 978-0-367-22124-9 (Volume 4) (hbk)
ISBN: 978-0-367-22127-0 (Volume 4) (pbk)
ISBN: 978-0-429-27339-1 (Volume 4) (ebk)

Publisher's Note
The publisher has gone to great lengths to ensure the quality of this reprint but points out that some imperfections in the original copies may be apparent.

Disclaimer
The publisher has made every effort to trace copyright holders and would welcome correspondence from those they have been unable to trace.

THE GREENING OF MACHIAVELLI

The Evolution Of International Environmental Politics

TONY BRENTON

THE ROYAL INSTITUTE OF
INTERNATIONAL AFFAIRS
Energy and Environmental Programme

EARTHSCAN
Earthscan Publications Ltd, London

First published in Great Britain in 1994 by
Earthscan Publications Ltd, 120 Pentonville Road, London N1 9JN and
Royal Institute of International Affairs, 10 St James's Square, London SW1Y 4LE

Distributed in North America by
The Brookings Institution, 1775 Massachusetts Avenue NW,
Washington DC 20036-2188

© Royal Institute of International Affairs, 1994

All rights reserved.

A catalogue record for this book is available from the British Library.

Paperback: ISBN 1 85383 211 1
Hardback: ISBN 1 85383 214 6

The Royal Institute of International Affairs is an independent body which promotes the rigorous study of international questions and does not express opinions of its own. The opinions expressed in this publication are the responsibility of the author.

Earthscan Publications Limited is an editorially independent subsidiary of Kogan Page Limited and publishes in association with the International Institute of Environment and Development and the World Wide Fund for Nature.

Printed in England by Clays Ltd, St Ives plc
Cover by Elaine Marriott

Contents

Acknowledgements ix
Preface xi
Summary and Conclusions xiii

1. Introduction 1
  1.1 The new agenda 1
  1.2 The environmental debate 2
  1.3 Aim and structure of this essay 6
  1.4 The role of the state 7
  1.5 Levels of response 10
  1.6 From Stockholm to Rio 12

2. The Birth of Environmentalism 15
  2.1 Prehistory 15
  2.2 Lift-off 18
  2.3 Causes 20
  2.4 The new environmentalism 24
  2.5 Disagreements 28
  2.6 Government responses 30

3. The Stockholm Conference 35
  3.1 The divergence 36
  3.2 Other issues 41
  3.3 The form of the conference 42
  3.4 The agreed texts 44
  3.5 The institutional outcome: UNEP 47
  3.6 Assessment 49

4. From Stockholm to Rio: Domestic Developments ... 51
- 4.1 Global trends ... 51
- 4.2 Developments in the OECD ... 53
- 4.3 The developing world ... 57
  - The nature of the problems ... 57
  - Population ... 60
  - Industrial pollution ... 64
- 4.4 The Soviet Union and Eastern Europe ... 71
- 4.5 Agriculture ... 75
- 4.6 Freedom and the environment ... 84

5. From Stockholm to Rio: International Developments ... 89
- 5.1 Overview ... 89
- 5.2 Marine pollution ... 91
- 5.3 Regional seas: the Mediterranean ... 95
- 5.4 The Convention on International Trade in Endangered Species (CITES) ... 100
- 5.5 Atmospheric pollution: acid rain ... 104
- 5.6 Environment in the European Community ... 107
- 5.7 Nuclear issues ... 115
- 5.8 Development of the North–South agenda ... 118
- 5.9 Conclusions ... 122

6. The Road to Rio ... 125
- 6.1 The second boom ... 125
- 6.2 Chernobyl ... 130
- 6.3 Toxic waste ... 131
- 6.4 Antarctica ... 133
- 6.5 The ozone layer ... 134
  - Background: before the 'hole' ... 134
  - The Vienna Convention ... 137
  - A change of momentum ... 138
  - The Montreal Protocol and London Amendments ... 140
  - Conclusions ... 143
- 6.6 The role of business and industry ... 146

6.7	Tropical forests	151
6.8	Moving up the international agenda	157

7. Climate Change 163
   7.1   The early work 163
   7.2   Climate change goes public 165
   7.3   Governmental responses 167
   7.4   Pressure on the US 171
   7.5   Target-setting 173
   7.6   The Intergovernmental Panel on Climate Change (IPCC) 177
   7.7   The developing countries join the debate 179
   7.8   The Second World Climate Conference 183
   7.9   The Convention negotiations: initial immobility 185
   7.10 New ideas, but no movement 187
   7.11 The ice breaks: the convention achieved 191
   7.12 Conclusions 193

8. Biodiversity 197
   8.1   The issue 197
   8.2   The negotiation 201
   8.3   The aftermath 204

9. UNCED: The Preparations 207
   9.1   The Global Environment Facility (GEF) 208
   9.2   UNCED structures and national objectives 210
   9.3   Agenda 21 and the Earth Charter 213
   9.4   Desertification and deforestation 215
   9.5   Institutions 217
   9.6   Finance and technology 219

10. UNCED: The Conference 223
   10.1 Atmospherics 223
   10.2 The five big issues 228
   10.3 The lessons 231

11. The Future **237**
   11.1 The challenge 237
   11.2 The benefits of prosperity and democracy 241
   11.3 Developing countries 244
   11.4 The global problems 247
          Biodiversity and forests 248
          Climate change 250
   11.5 Forces for international integration 252
          Negotiating processes 252
          Environmental science 254
          The NGOs 256
          Environmental altruism 258
   11.6 The environment and the international system 259
          The UN system 259
          International funding 261
          Trade 263
          A political centrepiece? 266
   11.7 Envoi: back to the roots 267

Index 273

Tables
5.1 Multilateral agreements with significant sections on the
    environment, 1950–1989 90
5.2 Key marine pollution agreements, 1954–1983 92

Figures
6.1 The structure of the UNCED negotiations 161

Acknowledgements

My first thanks must go to the Foreign and Commonwealth Office. I was the beneficiary in 1992/93 of their enlightened tradition of sending an officer each year to take up a fellowship at the Harvard Centre for International Affairs. It was in the course of that stimulating year that I did much of the research for this book. This does not mean that the Foreign Office, or the British Government, are in any way responsible for the views expressed here, which are entirely my own.

A number of individuals have been most generous with their time and attention. I must in particular mention Tony Fairclough, Gene Skolnikoff, Jacquie Roddick, Matthew Patterson, Fiona McConnell, Mike Harris, Richard Sandbrook, Diana Ponce-Nava, Gavin Watson, Ron Mitchell, Ted Parson, Naresh Singh and Ian Rowlands. And I have had unstinting help from Michael Grubb and Nicole Dando at the Royal Institute of International Affairs. The kindness of all these people has saved me from innumerable errors of fact and judgement. Those mistakes that remain are, of course, mine alone. I also owe a debt of gratitude to Gillian Bromley who did an excellent editorial job and to Hannah Doe who typeset the text.

Finally, my family – Sue, Tim, Kate and Jenny – have discovered how horrible it is to have a husband and father engaged on a project of this size. Without their tolerance and support it would simply not have been possible.

March 1994   Tony Brenton

Preface

It is not every day that a former government negotiator to the most complex environmental negotiations in history rings up and asks if we would be interested in seeing and considering a draft manuscript. It must be still rarer that the draft manuscript proves to fit in well with the Programme's research interests and to display a well-written combination of good coverage of the academic literature, practical experience and sometimes provocative judgement.

Such was our fortune with Tony Brenton. His book fills an important gap in the environmental literature: a pragmatic insider's account of how environmental issues have entered and developed on the stage of international politics, up to and including the first ever World Summit. It provides a highly readable account and analysis of a complex history, and Brenton adds many thoughts on the lessons that may be drawn from this experience of great relevance to assessing possible future developments.

The book completes a triad of publications on the broad sweep of international environmental developments by RIIA's Energy and Environmental Programme. In content and approach it provides a natural complement to the thematic analysis of Caroline Thomas' *The Environment in International Relations*; and a deep insight into the history that shaped the results of Rio, as summarized in our book *The Earth Summit Agreements: A Guide and Assessment*. Brenton's book completes a substantial contribution to the literature and to our understanding of international environmental affairs.

It was a privelege to be offered the project, and a pleasure to help Tony bring the book through the extensive process of review and refinement to the form published here.

March 1994
Dr Michael Grubb
Head, Energy and Environmental Programme

Summary and Conclusions

Although some environmental legislation (including a few international treaties) dates back to the last century, it was the great upsurge in the mid 1960s of Western popular concern about environmental pollution and the depletion of natural resources that gave birth to the modern environmental movement and pushed environmental issues high up on Western political agendas. Most Western governments established environment ministries and acted to tackle many pollution problems. In the international arena the 1967 *Torrey Canyon* accident initiated a pattern of disaster-driven international agreements, and marine pollution is the area in which the international environmental community learnt its craft.

These measures led up to the world's first major environment conference, in Stockholm in 1972. It became clear there that Western environmental alarm was not universally shared. The Communist bloc did not participate. The developing countries argued that their principal need was for economic growth, even if this consumed natural resources and caused pollution. The conference set the pattern for environmental negotiations in the years to come in the intensity of public and press interest it attracted and the openness of the negotiations. Yet for the most part the texts it produced were of little effect because of the divergent views which they had to accommodate. The most enduring formal product of the conference was the creation of the UN Environment Programme (UNEP) which was to act as midwife to the major environmental negotiations in the future.

Following Stockholm, government action in the West began to remedy some of the problems, and public concern fell back sharply. In the developing countries of the South, environmental problems intensified. Partly as a result of Stockholm, Southern governments began to introduce environmental legislation, although much of this has been imperfectly implemented. A growing number, in response more to domestic than international pressures, also introduced population policies, so that projected population growth (while still large) has been dramatically cut. In the communist states of Europe, the combination of centrally planned economies with authoritarian political systems produced extensive environmental degradation. Controlled markets also contributed to environmental damage in the field of agriculture, notably through overintensive production in the North and depletion of agricultural resources in the South. One overall lesson from the period is the close link between environmental protection, market economics and democratic political systems.

The years after Stockholm saw a growing volume of international environmental business. Much continued to focus on the seas, major milestones being agreements on the North Sea and the Mediterranean. The protection of endangered species was substantially reinforced through the CITES agreement which set precedents in demonstrating the effectiveness of using national self interest and trade measures to promote environmental ends. European countries, through a process of political and public pressure on those unwilling to move, reached agreements to tackle acid rain, despite the large clean-up costs involved. The European Community assembled an impressive array of shared environmental legislation and made environmental protection one of its central objectives. On other issues, such as nuclear safety, there was less international cooperation, and the West firmly resisted arguments in other fora that environmental pressures from the South justified large increases in aid.

The late 1980s saw a second boom in popular environmental alarm in the West prompted by a series of climatic and other disasters. Compared with 20 years earlier the concern focused more on pollution and less on resource consumption, and more on international issues and less on domestic ones. The explosion of the Chernobyl nuclear reactor led to significant international agreements on nuclear safety. Following some highly publicised incidents, developing countries demanded, and to some extent achieved, controls on the international trade in toxic waste. The agreements to phase out ozone-destroying chemicals were the largest and most groundbreaking of the period, and demonstrated the combined importance of public alarm (as aroused by the ozone 'hole'), international scientific consensus, and funding to assist developing country cooperation. It also displayed the way that sectors of industry are gradually identifying the environment as a source of potential profit rather than pure cost. At this time too, intense public concern prompted strong Western pressure (political and through aid) on Brazil and some other countries to reduce tropical deforestation. In 1987 the Brundtland Report made a big impact by linking environmental and developmental issues through the concept of 'sustainable development'. At the end of the decade, growing dissatisfaction with piecemeal approaches and the broadening of concern led to agreement to hold the Rio 'Earth Summit', and to launch negotiations on global climate change and the protection of biodiversity targeted to this deadline.

On climate, the Intergovernmental Panel on Climate Change established the scientific consensus necessary for serious negotiations on such an all encompassing and potentially expensive issue. The West split destructively on the issue of emission targets, and there were strong North/South differences about relative responsibilities and the amount and management of any aid made available to help developing country action. Given these difficulties, the climate convention was probably only achieved because of the need to have an agreement for Rio. The commitments it contains, on both aid and targets, are very limited but mark a politically significant first step in what could well become the biggest environmental negotiation ever.

The biodiversity negotiation was lower profile but just as fractious. The central issue was the insistence of developing countries that they should get a more substantial return from Northern exploitation (eg in drugs and agriculture) of Southern biodiversity. The North did not concede this, so the South (where most biodiversity is concentrated) was willing only to take on very limited conservation commitments. As with climate, the resulting agreement was a very small step towards dealing with the problem.

Another important strand was the establishment of the Global Environment Facility (GEF) as the main international instrument to finance action in developing countries on global environmental problems. The developed countries' ambition for a global forests convention was diluted to a very general statement of principles because of the insistence by some forest-intensive countries that this was a sovereign and not international topic, but political pressures did lead Rio to initiate negotiations on a desertification convention.

The main documents produced were the 'Rio Declaration' of principles for action, and 'Agenda 21', a global 'action plan' for the whole range of environment and development issues. The most contentious issue at the summit was international financial assistance; despite developing country hopes of big rise in global aid the final increase was very limited. It seems unlikely, given the diversity of approaches presented at Rio, that the concrete impact of the texts agreed there will be great, and it remains to be seen whether the UN Commission on Sustainable Development – established by Agenda 21 to monitor its implementation – will be sterilised by UN and North-South politics. Overall the outcome demonstrates that even global summits cannot achieve dramatic shifts in individual nations' policies. Nevertheless, the conference usefully focused global attention on environment/development issues, accelerated progress in key problems such as climate change, and underlined the high political prominence now attached to these issues.

Those countries with prosperous economies and democratic systems of government, such as in the West, are well equipped to tackle the environmental problems confronting them. Developing countries face greater difficulties but many of them can be expected to follow the Western trajectory (though there is a real problem over those countries which are already very poor and getting poorer). Individual countries are beginning to react but the challenge at the international level remains enormous. It remains unclear how fast progress is likely to be on the issues now at the top of the agenda; biodiversity (including deforestation) and climate. But despite the limitations of the Rio agreements, international environmental policy has a history of reaching destinations which seemed impossibly distant at the outset.

The past 20 years have seen growing forces for international cohesion, notably the 'process' style of environmental business under which commitments are ratcheted up over time, the growing sophistication of environmental science, the influence of the large and well funded environmental non-governmental organisations, and some indications that governments are ready on occasion to do more than is required by purely national interests. The international system too is adapting to environmental concerns through such developments as UNEP, the GEF, environmental pressures on the big UN Agencies and use of trade pressures for environmental ends. Occasional major events (such as Rio) help to focus attention on the issue, and some kind of permanent council might also contribute to the coherence of international activity. The outcome of Rio suggests, however, that the way forward may lie less in ambitious new international rules and agencies than in the fast growing practice of bringing home to individuals and individual countries the congruence between their private interests and the wider environmental good.

I have come across men of letters who have written history without taking part in public affairs, and politicians who have concerned themselves with producing events without thinking about them. I have observed that the first are always inclined to find general causes whereas the second, living in the midst of disconnected daily facts, are prone to imagine that everything is attributable to particular incidents, and that the wires they pull are the same as those that move the world. It is to be presumed that both are equally deceived.

ALEXIS DE TOCQUEVILLE

Chapter 1

Introduction

> *A man who wishes to act virtuously in every way necessarily comes to grief among so many who are not virtuous. Therefore if a prince wishes to maintain his rule he must learn how not to be virtuous.*
> NICOLÒ MACHIAVELLI

1.1 The new agenda

The purpose of this essay is modest, and largely personal. I worked on international environmental affairs, first in the European Community and then in the British Foreign Office, from 1986 to 1992. It was a good time to be there. I travelled the road from acid rain through the ozone layer to global climate change, taking in such picturesque detours as the North Sea, Chernobyl, and global deforestation.

For a conventional diplomat such as myself this was novel territory. Traditionally the business of diplomacy has been to manage the external relations of states. Internal affairs were rigidly excluded from consideration (indeed, the Charter of the United Nations explicitly prohibits the organization from interfering in the internal affairs of member states). But the environment was part of that fast-growing area of international business known at that time as the 'new international agenda' which also included such issues as human rights policy, drugs and AIDS. These are subjects on which internal policy and external policy are inextricably mixed. The solution of a domestic problem of, say, river pollution or rising drug addiction requires action at the international level; and international agreements on, say, human rights norms are explicitly intended to affect internal arrangements in the states involved.

The model of interaction among states was evolving from the collision of billiard balls, which touch only at a single point and do not change shape, to the mingling of immiscible oils on a glass surface where the whole shape and form of each component can be altered by the pressure of the components around it.

The emergence of this 'new agenda' of course reflects the enormous growth in recent decades of interdependence between states, and in particular of interdependence on matters other than the traditional political and security concerns that have hitherto dominated international relations. For example, over the past 40 years the proportion of the world's products traded internationally has more or less doubled and the absolute value of international trade has increased more than tenfold. Many more people are moving between countries (over the past 15 years the number of air miles travelled annually has tripled and the number of refugees has increased by a factor of six).[1] National currencies have largely drifted out of the control of their individual issuing governments and merged into a single unified global market with a turnover which is estimated at over $1 trillion per day.[2] A less widely advertised, but politically perhaps even more significant, area of growing international interpenetration is that of transborder data flows. The impacts of this phenomenon have so rapidly become global commonplaces that their novelty and long-term impact are in some danger of being forgotten. But such developments as, for example, the emergence of CNN, the ubiquity of *Yes Minister* on the world's one billion television sets, and the inability of the Chinese government to keep the news of Tienanmen Square, or the Soviet government the news of Chernobyl, from their respective peoples plainly have profound significance for the future shape and viability of particular national political regimes (notably those that have relied heavily on news management for their survival) and distinctive national cultures.

1.2 The environmental debate

The environment has formed a significant component of this new international interpenetration and interdependence. It is by now a truism, but also

---

[1] L. Brown, M. Kane and E. Ayres, *Vital Signs 1993-1994*, Earthscan, London 1993.
[2] P. Kennedy, *Preparing for the Twenty-first Century*, Random House, New York 1993.

Introduction

true, that many forms of pollution do not stop at frontiers. We will see below in some detail how growing human numbers and rising levels of economic activity have produced environmental consequences which are regional or even global in scale and which require international action to tackle them. The radiation from Chernobyl fell on 21 countries. Acid rain is now a continent-wide phenomenon in both Europe and North America. No single country is responsible for the decimation of the world's whale stocks, and only international cooperation can create the conditions for their regeneration. Climate change is global in both its sources and its impacts.

One striking aspect of this new subject matter is that, again unlike more traditional foreign policy subjects, it has been the source of intense domestic political controversy. In the UK, such issues as the Middle East dispute and the Falklands war caused relatively little domestic political argument. The dumping of nuclear waste and the control of power station emissions, on the other hand, have led to epic, and enduring, political rows. These disagreements, moreover, go beyond discussion of particular pollution issues to a quite profound ideological cleavage as to the true extent of the environmental threat that faces us, and the extent to which our lifestyles will have to change to meet it.

> **One striking aspect of this new subject matter is that unlike more traditional foreign policy subjects it has been the source of intense domestic political controversy. The dumping of nuclear waste and the control of power station emissions have led to epic, and enduring, political rows.**

One point of view, which has been espoused throughout Western Europe and the US by green movements and environmental non-governmental organizations (NGOs), holds that quite dramatic changes are needed in the resources we consume and the effluents we produce if human prospects are not to be seriously endangered by global environmental degradation. A persuasive array of statistics is offered to support this view. World population is increasing by about two million people a week; the world's tropical forest, which covers about the same area as the US, is diminishing by the area of Florida every year; one-third of the world's cultivated surface has been degraded to some degree by human agriculture; and the earth's climate is now changing at the fastest rate seen for 10,000 years. Jonathon Porritt, a leading British environmental campaigner, has written: 'The earth has just about coped with one billion people living a western, materialistic lifestyle. There's not a hope in hell that it will cope with five or six billion, let alone ten or eleven billion, subscribing to a similar fantasia'.[3] Such views cannot be dismissed as those of a radical and unrepresentative minority. The current Vice President of the US (not a position one generally reaches by being radical and unrepresentative) has, for example, called for the rescue of the environment to become 'the central organizing principle for civilization.'[4]

This school of thought has taken up a haunting parable, put into circulation by Garret Harding (who ascribes it to an earlier author), known as 'the tragedy of the commons'.[5] The image is of a group of herdsmen grazing their cattle on common land. All know that the addition of an extra cow to any herd increases the pressure on the common. But that cost is shared among all the herdsmen while the profit from keeping the extra animal accrues to its owner alone. Thus each herdsman gains from every extra animal he keeps, and will add to his herd. With Sophoclean inevitability the herds expand, the common is overexploited, its agricultural worth exhausted, and the way of life of the herdsmen themselves destroyed. The application of this story to national 'commons' of, for example, clean rivers and pure air, and to the 'global commons', such as the oceans and the global atmosphere, is plain. Left to themselves, profit-maximizing individual users (in the nation-state case) and self-interested nation states (in the international case) will expand economic activity, knowing that the consequent polluting emissions will be dispersed among all, while the economic benefit from the extra production will be confined to the producer alone. Thus, inevitably, emissions will rise and the environment in which they are deposited will degrade, ultimately to the point where our prosperity, and even survival, may be at risk.[6]

---

[3] J. Porritt in J. Porritt, ed. *Save the Earth*, Turner Publications, London 1991.
[4] A. Gore, *Earth in the Balance*, Houghton Mifflin, New York 1992.
[5] G. Harding: 'The Tragedy of the Commons', *Science*, 162, 1968 pp 1243-8.
[6] I use the word 'parable' deliberately. Much of the rest of this book is intended precisely to explore the question of whether the earth really faces a tragedy of the commons. A number of authors, notably E.E. Ostrom (*Governing the Commons*, Cambridge University Press, Cambridge 1990) have pointed out that even at the primitive local level epitomized by Hardin's herdsmen there are cases, e.g. of common meadows in Switzerland and shared

In its international application this forecast of environmental destruction of course draws heavily on a tradition in political thought most popularly associated with the name of Machiavelli. It was he who urged 'the Prince' (in contemporary terms, the nation-state government) amorally to pursue his own interests, and expediently to set aside any concern for the global good. In fifteenth-century Italy the product of such politics was an era of warfare and betrayal among petty statelets, which accordingly neglected the very real external dangers they jointly faced until they were consumed by them.

On the other side of the environmental argument, John Maddox, editor of *Nature*, has written of the 'pantomimic wave of overreaction to some of the supposed dangers of environmental contamination'. This view has been widely echoed in sober and intelligent newspapers such as the *Wall Street Journal* and the London *Economist*. Indeed, given the extent of public concern about environmental issues, it is striking how dismissive well informed and thoughtful people remain about the whole environmentalist case. They will accept that particular incontrovertible instances of pollution and its effects such as London's 'killer' smogs, oceanic oil slicks, or even the hole in the ozone layer, require urgent action. But they are much more sceptical about the general thesis that a radical change in relationship between mankind and nature is required if disaster is to be avoided. They point out that the environmentalists' trend lines to disaster take no account of the technological and other developments, which have regularly falsified such predictions in the past, for example in the case of Malthus. The gloomy predictions are flatly contradicted by experience in the West, where in many ways the environment is getting not worse but better, and where government action to tackle particular environmental problems is a regular event. Indeed, many will argue that political concern on the issue is grotesquely out of proportion to its true importance, partly because of sensationalist media coverage. During the late 1980s, when popular environmental concern was approaching its peak in the UK and throughout the West, I regularly heard such doubts about the objective justification for the popular frenzy advanced at high-level meetings on the subject, both inside and outside government.

---

water rights in Spain, where the users have managed to establish arrangements to avoid the overuse and ultimate exhaustion of the resource.

1.3 Aim and Structure of this essay

This essay is an attempt to chart a course between these two points of view. At the time I was directly dealing with the subject matter I was uneasily aware of the tension but had no leisure to explore the issues in any depth. A sabbatical year from the Foreign Office, spent at Harvard, gave me that leisure. The central theme which I have tried to address is the extent to which environmental politics, particularly at the international level, has produced an adequate response to the environmental challenge. Are national governments and the international system doing enough to tackle the environmental problems that confront us?

This is plainly a question with a scientific component. But it is by no means exclusively, or even predominantly, a scientific question. The simple model according to which scientists identify a threat and governments act to solve it has emphatically not been the pattern of international environmental activity over the past 20 years. Even when scientists are agreed on the nature of the problem (and on many environmental issues, especially the large ones such as climate change, they are often far from being so), the route to remedial action is a long one. Governments have many other considerations, notably of economic impact and domestic public acceptability, to take into account in framing their responses. Thus, while I have included some description of the science relating to the various issues where that has seemed necessary, the bulk of what follows is devoted to a history of international environmental issues from their first great eruption on to the political scene in the 1960s up to the present day. I have not endeavoured to be encyclopedic, but simply to cover the principal waystations with some explanations of their significance. Like most other items on the 'new international agenda' much of the international activity is incomprehensible without some understanding of domestic environmental politics, which I have also, therefore covered at some length.

The history, and the implications drawn from it, are intended to be objective, and indeed my own views have shifted (in a markedly more optimistic direction) in the course of the research that has produced this book. Nevertheless, on territory so strongly ideologically disputed objectivity is a slippery concept (some would say a chimera) and it is therefore only fair to readers to forewarn them of where I stand. First, I prefer to follow Machiavelli in avoiding judgements as to whether the world's current economic and political ar-

Introduction   7

rangements are 'right' or 'wrong'. I am more interested in the practical question of how, given these arrangements, we can best tackle the environmental problems that face us. Second, I think of myself as a moderate (or, in current parlance, 'light green') environmentalist. It is difficult to work for any period of time in the field without being persuaded that the noxious byproducts of industrialization and economic development constitute a direct threat to future human health and prosperity unless vigorous action is taken to contain them. Moreover, this threat is now of sufficient magnitude to constitute one of the central challenges currently facing national governments and the international system. I am not, however, persuaded that we yet face an environmental crisis which will overwhelm us unless we make very rapid and dramatic changes to our lifestyles and aspirations. Third, I am an unreconstructed anthropocentrist. I can see the importance of other species and the ecological balance for the survival and amenity of mankind. But I see no basis for 'dark green' environmentalist contentions that they have any intrinsic value over and above this. If a starving sailor were trapped on an island with the last dodo, I would support his right to eat the dodo for dinner.

1.4 The role of the state

The emphasis of this book is very much upon the action of states. This may seem paradoxical. The emergence of the 'new agenda' is plainly a symptom of the diminishing autonomy of the individual nation state. Governments are decreasingly able to set their own exchange rates, control population movements across their borders or prescribe what television programmes their people should watch. It is therefore worth asking whether a history of international environmental action which focuses on government policies and attitudes doesn't miss the point. Maybe the real authority lies, or is coming to lie, with the more transnational players, international business, the NGOs, the scientific community and the multinational organizations? It will become clear below that these other organisms have indeed come to play a significant part in international environmental business as it is currently conducted. But that part has been either (in the case of NGOs, scientists and business) as lobbyists of national governments or (in the case of multinational organizations) as their agents. This is not to say that they have lacked influence. We will be looking at cases where the public or private lobbying activities of NGOs, business or the scientific community have significantly affected the course and outcome of particular negotiations. But the centres of decision remain in national capitals. It is my firm judgement that over the period reviewed in this book the key players in international environmental matters have remained the national governments.[7] Certainly their policies adjust to pressures from business, NGOs and science, but they are also subject to a host of other domestic political pressures, often remote from the environmental issue ostensibly under discussion, which can play a key part in shaping the eventual outcome of a negotiation.

One other point about the nation state of which it is easy to lose sight in books about international politics is that it is not monolithic. Phrases like 'the UK decided...' and 'Brazil was persuaded...', which I use a lot below, are in fact shorthand for the outcome of the arguments between the various interest groups and other political players within the state concerned. It is important to bear this in mind because it places the role of international discussion of 'new agenda' issues in its proper context. The most important impact of such discussion is to strengthen the hand of particular domestic factions within the states concerned. The role of the GATT process in helping free traders in national administrations to resist the pressures from protectionist interest groups, partly on the basis that similar interest groups will be similarly resisted in other countries, is a conspicuous example. Similarly, a lot of the international environmental action which we will examine must be seen for what it is: different national environmental ministries and lobbies establishing common cause in international fora so as better to be able to overcome counter-environmental pressures within their own administrations.

The non-monolithic quality of the state is important for our story in another respect too. In many developing countries (and in some developed ones as well) the capacity of the central government to impose its will on its domestic public is extremely limited. It has been suggested that 'in many, if not most, of the modernising countries of Asia, Africa and Latin America... there is a shortage of effective, authoritative, legitimate government'.[8] This shortage,

---

[7] The same of course remains true in areas of international cooperation much more advanced than that on the environment, such as the European Community.

[8] S. Huntingdon, *Political Order in Changing Societies*, Yale University Press, New Haven 1968.

Introduction

as manifested for example in the inability of the Brazilian authorities to impose effective controls on forest burning in Amazonia or the inability of the Indian government to stamp out the local corruption which makes a mockery of many of its environmental regulations, plainly has serious implications for the capacity of developing-country governments both to tackle their own domestic environmental problems and to participate in wider efforts to deal with international issues.

This difference in governmental authority is just part of the huge economic, political and social gulf which lies between the way of life of the developed, industrialized 'North' and that of the underdeveloped and still heavily agricultural 'South'. It is unsurprising therefore that a substantial part of our analysis is framed in North-South terms. This is of course a crude generalization. Neither the developed nor the developing countries form a homogeneous bloc. Economic, political and ideological differences among developing countries are at least as wide as those between North and South. A rapidly industrializing developing country such as Mexico in many ways has more in common with an OECD state like Portugal than with one of the least developed like Somalia. With the burst of economic growth in east Asia, and incipiently in Latin America, the line is becoming harder and harder to draw. Nevertheless it is there. The objective gap in terms of wealth, style of government, social organization and myriad other factors including impact on the global environment between the majority of developed and majority of developing countries is still vast. This cleavage between North and South dominated the political background to most of the environmental business of the period studied here (not least because it dominates so much of the business of the UN). Any history must reflect that fact.

> **The gap in terms of wealth, style of government, social organization and myriad other factors between the majority of developed and majority of developing countries is still vast. This cleavage between North and South dominated the political background to most of the environmental business of the period studied here (not least because it dominates so much of the business of the UN).**

This North-South divergence also runs through our story in the form of continuing tension between 'environment' and 'development'. These two subject areas can be viewed as the two faces of a larger general question which might grandiloquently be described as being concerned with 'the future amenity of the planet'. As we will see in detail below, Northern populations and politicians have for the most part been confronted with the 'future amenity' problem from an environmental perspective. Will there be enough global resources to maintain our standard of living? Can we tackle the polluting emissions which threaten that standard of living? How will growing populations impact upon our prosperity? And so on. The Southern angle on the 'future amenity' question has been focused much more on development than on environment. How can we tackle hunger, ignorance and disease and achieve a decent standard of living for our people? How can we begin to offer them life chances comparable to those which are routine in the prosperous North? That these are two aspects of the same problem is clear from the rather trivial observation that levels of global pollution and resource depletion are plainly a function of levels of global consumption. The more the world develops, the more its environment is apparently threatened. This essay, being written by a Northerner, approaches these tensions from a Northern perspective. Our way into the problems we discuss is through their environmental manifestations. It is important however to recognize, as I have endeavoured to do below, the centrality of the 'developmental' perspective for the South and the consequent need to accommodate that perspective if global agreements on 'future amenity' issues are to be achieved and made to work.

1.5 Levels of response

Much of the discussion in following chapters concerns the negotiation of various national and international instruments on the environment-laws, statements, treaties, protocols and so on. In order to chart a way through this bewildering forest of texts I have found it useful to distinguish three 'levels of response' to environmental problems. These may be applied in both the national and international contexts. At the top, 'response level three' is the situation where a law (in the national case) or a treaty (in the international case) has been agreed and entered into force and has contributed to changes of

Introduction

behaviour by individuals, or of policy by nations, which are actually ameliorating the problem[9]. Examples examined below include the UK's Clean Air Acts, which have eliminated London's 'killer' smogs, and the European Community's acid rain legislation which has helped bring about a cut in European sulphur dioxide emissions. The next step down, 'response level two', applies where a legally binding instrument, a national law or international treaty, has entered into force but has not affected actual behaviour, perhaps because it is too recent or because there is no mechanism for its enforcement. Examples include a large proportion of developing country and communist bloc environmental legislation which has never been effectively implemented.

Finally there is 'response level one', which applies where governments or groups of nations have agreed non-legally-binding texts such as conference statements but have not (yet) moved on to enact laws or adopt treaties to tackle the problem under discussion. Examples of level one texts include the UK's regular environment White Papers, which set out legislative and other plans for national environmental protection, and the Earth Summit's 'Agenda 21', which purports to do the same at the world level.

It is very easy to be dismissive of level one responses. It is of course far easier for governments to talk firmly about environmental problems than to take the (often expensive and politically difficult) action required to tackle them, and it is frequently very difficult to tell, when responses are at level one, whether governments seriously intend to tackle the issue or are simply swimming with a tide of popular opinion. We will see below numerous examples of the difficulty the international community has had in moving from a level one response to level two and level three responses to particular problems. Commentators have indeed lamented that on environmental issues 'the

---

[9] Treaties go through a number of distinct stages before becoming part of international law. First they are negotiated by interested countries. Then those countries which are content with the outcome of the negotiation sign the resulting document. Those that have signed then consider whether to ratify – i.e. accept the obligations of the treaty as binding. Once a certain number (specified in the treaty itself) have ratified, the treaty 'enters into force' and becomes binding on those that have ratified together with those who do so later. There still remains the question, of course, of whether those who have committed themselves in the treaty actually observe it. We will see below examples of treaties which have fallen at various of these fences and thus never become operational.

ratio of words to action is weighted too heavily towards the former'[10]. It is however possible to take this scepticism too far. In my experience, governments are very conscious that once they begin to commit themselves to a line of environmental action at level one the pressures (from public opinion, special interest groups and other governments) for them to move forward to levels two and three become much harder to resist. That is why governments put so much effort into protecting their positions even at level one. Words may be cheap but they are not free.

There is, in fact, an obvious measure of the true significance of a level one text as an indicator of future action: its particularity. Those declarations (such as the various utterances of the '30% Club': see Chapter 5 below) which contain precise numbers, dates and objectives are very difficult for the signatories subsequently to evade or ignore. If, on the other hand, an international statement is long on worthy generality but short on concrete undertakings, that is a clear sign that the governments concerned can agree that virtue is good but not on what precisely they are going to do to promote it. This distinction will be noted a number of times in what follows.

> Those declarations which contain precise numbers, dates and objectives are very difficult for the signatories subsequently to evade or ignore. If, on the other hand, an international statement is long on worthy generality but short on concrete undertakings, that is a clear sign that the governments concerned can agree that virtue is good but not on what precisely they are going to do to promote it.

1.6 From Stockholm to Rio

Happily for the chronicler, the period upon which most of this essay concentrates is framed by the two biggest international environmental events that have ever taken place: the UN Conference on the Human Environment in Stockholm in 1972, and the UN Conference on Environment and Develop-

---

[10] M. Holdgate, M. Kassas and G. White eds., *The World Environment 1972-82*, Tycooly International, Dublin 1982.

ment in Rio de Janiero in 1992. Each of these two conferences provides an excellent snapshot of the state of global environmental attitudes at the time it took place, and together they offer a rather clear view of what changed, and did not change, in the twenty years between them, in particular, how the environment rose in a mere quarter-century from obscurity to become the subject of the first ever world summit. As the story approaches the present so it becomes significantly more detailed, both because the forces at work become increasingly relevant to how things are likely to go in the future, and because I find it easier to write, having, for the most part, been there.

The book is structured as follows. Chapter 2 summarizes the 'prehistory' of international environmental action leading up to the first great explosion of popular environmental concern in the West in the 1960s which gave birth to the modern environmental movement. Chapter 3 looks at the first great international manifestation of this new environmental concern, the Stockholm Conference of 1972. Chapter 4 then looks at the evolution of the domestic politics of the environment in individual nations in the years following Stockholm, with sections on national population policies, on the environmental implications of national agricultural policies and on what the history of national environmental action has to teach us about the relationship between environmental protection and political freedom. Chapter 5 looks at international developments over the same period focusing both on the major growth areas of marine pollution, acid rain, endangered species and the environment policy of the European Community, and on areas where international action might have been expected to play a larger role than it did, such as desertification, nuclear matters and the whole question of North-South economic (including environmental) relations. Chapter 6 then examines the second great boom of Western environmental concern in the late 1980s and the major accompanying international events, notably the response to the Chernobyl accident and the negotiations on the future of Antarctica, international toxic waste management and the ozone layer, the latter leading naturally to a more general discussion of the changing attitude of business and industry to environmental issues. This brings us to the 1990s and the big environmental negotiations of that time. Chapter 7 is devoted to climate change and Chapter 8 to biodiversity, while Chapters 9 and 10 summarize the political peak, so far, of international environmental history: the Rio Summit.

But the aim of the analysis is not primarily historical. It has been argued, especially since the end of the Cold War, that the problems of the global environment are now the most demanding test facing the international political order. It is quite conceivable that the negotiations on global climate change could turn into the largest, in terms of their total economic impact, that mankind has ever undertaken. My central aim in looking at the history is to gain some insight into how effective international environmental action is likely to be in the future. There are big questions here. Is the international system as at present constituted capable of responding to the threat of global environmental degradation? What changes to existing international mechanisms might make it more able to do so? Chapter 11 is addressed to these questions. Evidently the answers must to a significant degree be unsure and speculative. We are after all shooting at not one but two moving targets. Our understanding of environmental issues is in rapid evolution, as are the systems we have developed to deal with them. But even at the level of speculation it must be worth reflecting on the question whether, Machiavelli notwithstanding, the human race has the political capacity to avoid the tragedy of the commons.

Chapter 2

The Birth of Environmentalism

> *There is one thing stronger than all the armies in the world and that is an idea whose time has come.*
> VICTOR HUGO

2.1 Prehistory

According to the Oxford dictionary the words 'ecology' and 'environment' first took on their modern meanings in the early 1960s. The words existed before, but with scientific definitions. It was in 1963 that Aldous Huxley gave the word 'ecology' its current sense in a paper called 'The Politics of Ecology' – a title which, however banal it may seem today, was linguistically innovative then. At this time, too, 'environmentalism' ceased to connote a scientific theory about the relative weights of nature and nurture and entered common parlance as a major new force in public affairs.

This does not mean that there was no action on what we now describe as environmental issues before the 1960s, only that there was no word for it. Individual countries in Western Europe and North America were already tackling particular conservation and pollution issues by the end of the nineteenth century: perhaps the first major antipollution action of the industrial era was the UK's Alkali Act of 1863.[1] The history of cooperative action by groups of countries to tackle 'environmental' problems whose effects extended across international borders goes back almost as far. It was in 1872 that the Swiss first proposed the establishment of an international commission to protect migrating birds. Probably the first ever international environmental agree-

---

[1] J. McCormick, *The Global Environmental Movement*, Belhaven Press, London 1989.

ment was the 1900 Convention for the Preservation of Animals, Birds and Fish in Africa, signed in London by the European colonial powers with the (strikingly farsighted) intention of preserving game in east Africa by limiting ivory exports from the region.[2] The late nineteenth and early twentieth century also saw an international convention to protect fur seals, an agreement among littoral states on the management of the Rhine and a US-Canadian agreement on the protection of migrating birds. The newly born clutch of international organizations of the first half of the twentieth century often took on environmental responsibilities: the Food and Agriculture Organization (FAO) for conservation of natural resources, the International Labour Organization (ILO) for worker protection against occupational environmental hazards and the International Maritime Organization (IMO) – at that time the Intergovernmental Maritime Consultation Organization (IMCO) – for marine pollution control. The first effort at global environmental management may date from 1909 with an (unsuccessful) US initiative to convene a world conference on natural resource conservation. Probably the first global resource management instrument actually agreed was the whaling convention of 1931, which led to the establishment of the International Whaling Commission in 1946.

The most striking feature of this prehistory of international environmental activity is its emphasis on conservation and wildlife problems. This stemmed from the early rise of nature preservation movements in the UK and US (although action on these issues was not entirely confined to developed countries; the 1940 Convention on Nature Protection and Wildlife Conservation in the Western Hemisphere included 14 developing country signatories).[3] A very early example of international resolution of a pollution issue was the 'Trail Smelter' case of 1935 in which, following the adjudication of a specially established international arbitration tribunal, Canada paid compensation to the US for damage to farm crops caused by the emissions of a metal refinery in British Columbia.[4] The first pollution problem to receive extended attention at the international level was the discharge of oil from tankers into

---

[2] Ibid.
[3] Ibid.
[4] L. Caldwell, *International Environment Policy, Emergence and Dimensions*, 2nd edn, Duke University Press, Durham, NC and London 1990.

the oceans. This was already a source of concern in the 1920s, largely because of its effects on birds and beaches. After a number of failed international initiatives on the subject the British government, prompted by special interest groups such as bird protection societies, convened a meeting in London in 1954 which agreed upon the first ever international instrument to tackle pollution: the International Convention for the Prevention of Oil Pollution.

This agreement was prophetic in a number of ways.[5] Its principal negotiators and early signatories were the developed countries of the North Atlantic, with developing countries joining much more slowly. A key motivation driving the parties to seek progress through international agreement rather than domestic legislation was their determination that their tanker fleets should not be placed at a competitive disadvantage by being subjected to tougher domestic regulation than that imposed by other countries on their fleets. Over time, improved knowledge of the problem and developing technology permitted the agreement to be strengthened through regular amending conferences. This process was helped by the emergence of a technically competent and politically low-profile international organization which could host the negotiations (the IMCO, later to become the IMO, which was established as a UN specialized agency in 1958), as well as of an experienced international community of negotiators and experts to carry the process forward. An important constraint on the process, and on the pace of ratification of the agreement, with its various subsequent amendments, was the speed with which tanker owners were willing to introduce new and less polluting technology into their fleets. This limitation was exacerbated by the failure of the parties to agree on any effective system for surveillance of violations, so that implementation was dependent on the pace of introduction of the new technology which was as slow as the tanker owners could make it. The agreement was nevertheless strengthened by rounds of amendments in 1962, 1969 and 1971. Finally, the statistics suggest that it did indeed make a significant contribution to the reduction of deliberate oil discharges into the oceans. A 1981 IMO report concluded that such discharges fell by about 30% over the 1970s, a period when the amount of oil transported by sea increased by about 17%. Thus this

---

[5] Accounts of the agreement and its background are to be found in Churchill and Lowe, *The Law of the Sea*, 2nd ed, Manchester University Press, Manchester 1988; E. Gold, *Handbook on Marine Pollution*, Assuranceforingen Gard, Rotterdam 1985.

agreement with its various additions constitutes our first example of the international community achieving a level three response to an environmental pollution problem.

The oil agreements were very much the product of a government – and expert – driven process which attracted little public interest except among special interest groups like the Royal Society for the Protection of Birds. Public concern played a much larger part in the other international agreement of the period which had a significant environmental component: the Partial Nuclear Test Ban Treaty of 1963. But while public interest in this treaty was to some extent motivated by growing fears about the effects of radioactive fallout (in US opinion polls in 1963 the proportion of respondents offering concern about fallout as a justification for the treaty ranged between 12% and 21%), this was clearly only a subsidiary consideration in the minds of both publics and governments by comparison with the military and security implications.[6]

2.2 Lift-off

It is probably something of a coincidence, therefore, that it was in this period, the early 1960s, that the genuinely new phenomenon of widespread public fears on the general subject of pollution emerged, and, as noted above, the concept of 'environmentalism' entered the language. The catalysing, or at least emblematic, event which is often taken as marking the birth of this new environmental consciousness was the publication in 1962 of *Silent Spring*,[7] Rachel Carson's eloquent indictment of excessive pesticide use. This book stands at the head of, and has in many cases been the inspiration for, the long stream of environmentalist literature which has followed it. Quite clearly it appeared at a moment when the public was ready for what it had to say. It sold half a million copies in hardback, was on the US bestseller list for 31 weeks (a feat perhaps helped by the counter attack it elicited from the chemical companies) and was swiftly published in 15 other countries. It unleashed

---

[6] See eg US Arms Control and Disarmament Agency, *Arms Control and Disarmament Agreements*, Washington, DC August 1980; Jacobson and Stein, *Diplomats, Scientists and Politicians: The US and Nuclear Test Ban Negotiations*, University of Michigan 1966.
[7] Rachel Carson, *Silent Spring*, Houghton Mifflin, Boston 1962.

a flood tide, first in the US and then more widely in the West, of debate and writing which swiftly extended beyond the issue of pesticides to the whole question of what mankind was doing to the natural environment.

Once incorporated into the vocabulary, 'the environment' moved with extraordinary rapidity up the agenda of public concern. In the US the number of opinion poll respondents who identified pollution as among the most important problems for government quadrupled between 1965 and 1970.[8] Membership of the Sierra Club and the National Audubon Society, until that time rather staid conservation organizations with a steady few tens of thousands of members each, leapt by the early 1970s to 140,000 and 200,000 respectively.[9] The period also saw the establishment of a spate of new, and in general much more radical (or at least litigious), environmental groups, notably the Environment Defense Fund (1967) and the Natural Resources Defense Council (1970). The Sierra Club itself split along the fault-line between the traditional conservationism and the new environmentalism and thus gave birth to the Friends of the Earth. With the establishment of the World Wide Fund for Nature, Friends of the Earth and Greenpeace, the world saw its first *international* environmental pressure groups operating in a number of countries with close links between their various national organizations. The climactic coming of age of the environment movement in the US took place on 'Earth Day', 22 April 1970 in which 20 million people participated and which provoked *Time* magazine to dub the environment 'issue of the year'.

> **The publication in 1962 of Silent Spring unleashed a flood tide of debate and writing which swiftly extended beyond the issue of pesticides to the whole question of what mankind was doing to the natural environment.**

The US pattern was broadly repeated throughout the developed world. Studies on France, Germany, the UK, Sweden, Switzerland and the Netherlands all showed a significant rise in public concern about environmental issues

---

[8] J. Springer and E. Constantini, 'Public Opinion and the Environment: An Issue in Search of a Home', in Nagel, ed., *Environmental Politics*, Praeger, New York 1974.
[9] Ibid.

through the 1960s and early 1970s.[10] The first national 'Green' party was established in New Zealand in 1972. The case of Japan is particularly interesting. In a polity with a well-earned reputation for public passivity and conformity, the first ever explosion of citizen activism was driven by concern about environmental pollution. By 1973, about 3,000 environmentally concerned citizens' movements had come into existence. The number of pollution complaints received by Japanese local government was 75,000 in 1971 – double the number received in 1969 and four times the number in 1966.[11]

2.3 Causes

It has been suggested that 'when the history of the twentieth century is finally written the single most important social movement of the period will be judged to be environmentalism'.[12] Certainly it is difficult to identify another public issue which achieved so high a level of popular concern from such a low base with such speed and in so many countries as did the environment in the late 1960s. What were the factors that came together to produce this result?

First and foremost, pollution had in fact been rising.[13] The period since the Second World War had seen an unprecedented growth of material wealth. Gross world product more than doubled between 1950 and 1970; and a lot of this growth took place in highly effluent industries. US production of synthetic chemicals rose five fold between 1950 and 1970; world sulphur dioxide emissions rose by 50% over the same period, while world consumption of

---

[10] H. Kitschelt, 'Explaining Contemporary Social Movements: A Comparison of Theories', paper delivered at the August 1989 annual meeting of the American Political Science Association, Atlanta, Georgia.
[11] E. Krauss and B. Simcock, 'Citizens' Movements: The Growth and Impact of Environmental Protest in Japan', in K. Steiner, E. Krauss and S. Flanagan eds, *Political Opposition and Local Politics in Japan*, Princeton University Press, Princeton NJ, 1980.
[12] R. Nisbet, *Prejudices: A Philosophical Dictionary*, Harvard University Press, Harvard 1982.
[13] The figures that follow are drawn from P. Haas, *Saving the Mediterranean*, Columbia University Press, New York 1990; *World Resources 1992-93*; a report by the World Resources Institute in collaboration with the United Nations Environment Programme and the United Nations Development Programme, Oxford University Press, Oxford 1992 and M. Tolba et al., *The World Environment 1972-1992*, Chapman and Hall, London 1992.

fossil fuels more than doubled, and emissions of toxic heavy metals rose proportionately. The disposal of toxic waste became a major industry – for example, over 20,000 tonnes of nuclear waste were dumped at sea in 1970, with a measurable effect on global marine radioactivity. World population increased by about 40% (1 billion people) between 1950 and 1970, thus causing anxiety about whether the planet could support, or tolerate the pollution from, the exponentially rising number of people living on it. Accessible bodies of water such as Lake Erie and the Mediterranean Sea were pronounced (respectively) 'dead' and 'sick' because of the volumes of pollution they were having to absorb and the destruction of marine life this was causing.

But it is also clear that the rise in environmental degradation does not by itself account for the very sharp growth in public awareness of it. It has rightly been pointed out that 'the change in public attitudes has been much faster than any change in the environment itself'.[14] Moreover, by the time public concern reached its maximum intensity (about 1970) some key forms of pollution were well on the way to being tackled. As a result of the 1956 UK Clean Air Act, London had seen an 80% reduction in smoke, a 40% reduction in sulphur dioxide emissions and doubled hours of winter sunshine. The Thames, which in 1960 was too polluted to support any fish life, in 1970 was host to more than 40 species. Carbon monoxide and sulphur dioxide levels were falling in most major US cities. Oregon's most polluted river system was well on the way back to health.

One clear additional reason why the public attitude shifted as abruptly as it did is the intensity with which the Western press took up the issue. In the US context it has been observed that 'not since the Japanese attack on Pearl Harbor has any public issue received such massive support in all the news media'.[15] A series of surveys of various sectors of the US press suggests that coverage of the environment moved from eighth among 14 major issues in the 1960s to third in 1970 to become the top domestic issue (in terms of column inches) by 1973.[16] This quantity of press coverage was in fact boosted by a

---

[14] A. Downs, 'Up and Down with Ecology: The Issue/Attention Cycle', *The Public Interest*, 28, Summer 1972.
[15] H. Sprout, 'The Environmental Crisis in the Context of American Politics', in Roos, ed. *The Politics of Ecocide*, Holt, Rinehart and Wilson, New York 1971.
[16] Springer and Constantini, 'Public Opinion and the Environment'.

number of significant disasters, which, as any journalist will confirm, always make good copy. In 1959 there was a major case of mercury poisoning near Minamata Bay in Japan; in 1967 the supertanker *Torrey Canyon* ran aground, causing a massive oil spillage in the English Channel; and in 1968 the Japanese authorities finally recognized the widespread and highly degenerative 'Itai Itai disease' in Toyama prefecture as resulting from cadmium pollution.

Three other causes for the boom in Western environmentalism may be identified. The first is the growing level of material affluence in developed economies through the 1960s, which seems to have produced a shift in preferences from increased material consumption to improvements in non-material factors of life, such as environmental quality.[17] This explanation is particularly persuasive in the case of Japan, where the 1960s saw the benefits of the post-war economic miracle feed through to the population, so generating levels of prosperity at which it became possible to contemplate alternatives to the 'growth at all costs' philosophy which had prevailed up to that point. But one can see the same factors, in a slightly less stark form, operating all across the developed world where the environmentalist movement broadly took a posture hostile to economic growth and nevertheless enjoyed significant public support.

> Three other causes for the boom in Western environmentalism may be identified: the growing level of material affluence, the counter-culture and the 'shrinking planet'.

The second and closely related factor is that the 1960s were, of course, the era when all across the developed world the counter-culture became part of the mainstream. The decade saw a youthful rebellion against 'establishment' values, and in particular against their more materialistic aspects. The hold of those values was substantially weakened, and new issues, of which the environment was a prominent example, were thrust forward as important topics for public attention and action. Thus, as a British commentator rather tartly observed, 'in the United States groups of students tired of burning down their colleges in the name of peace had decided to burn them down in the name of

---

[17] McCormick, *Global Environmental Movement*, notes similar, if less dramatic, shifts in public mood associated with periods of prosperity in the UK in the 1890s, 1920s and 1950s.

conservation'.[18] This shift in social values took place to different extents in different countries, perhaps greatest in Scandinavia and least in Japan; but there is little doubt that it happened everywhere, and increased environmental concern was part of it.

The third factor which helped boost environmental awareness in this period was the 'shrinking planet'. The 1960s saw the start of the revolution in communications and global economic interdependence which continues today. Suddenly Japan with its mercury pollution and China with its teeming millions were just next door, and more particularly on the television every night. Suddenly the diseased fishermen of Minamata, the dying lakes of Scandinavia and the poisoned rivers of Oregon could be presented as part of a global syndrome. Many of the emerging environmental issues were intrinsically international in nature – the oceans, migratory wildlife, the global atmosphere. The reality of growing global interdependence brought with it increased public awareness of events outside one's own national frontiers, and hence increased public attention to issues of international scope, of which the environment became a prominent example. Again, this phenomenon occurred to different degrees in different countries, going furthest in Western Europe, where already a group of relatively small and closely linked countries had recognized their growing interdependence and were responding through the development of the European Community, and least far in the US, whose size and geographical situation still left the rest of the world looking pretty distant. But even in the case of the US the outside world was significantly less remote than before, as is clearly demonstrated by the proportion of press coverage devoted to international developments in these years. Perhaps the most potent icon of this new 'global consciousness' has been the 'earthrise' photograph taken by *Apollo 11* in 1969, and extensively exploited since by exponents of the 'fragile planet' view of the human enterprise.

One is nevertheless left with the feeling that the speed and the intensity with which Western environmental concern emerged in the 1960s are greater than can be explained by the mundane factors listed above. In the search for historical precedents one is tempted to recall the waves of apocalyptic frenzy

---

[18] B. Levin, *The Pendulum Years: Britain and the 60s*, Jonathan Cape, London 1970. For a good account of the whole period, see D. Caute, *'68: The Year of the Barricades*, Hamish Hamilton, London 1988.

which hit medieval Europe from the eleventh century on onwards. Then too a period of social and economic dislocation produced a series of outbreaks of popular fervour whose tenor was that modern man had gone astray and that the way to salvation lay through renunciation of materialism and espousal of a simpler, 'unpolluted' way of life.[19] More prosaically, modern sociology offers the model of the 'issue/attention cycle' by which a public issue, such as ecology, for some reason captures popular attention and then becomes the object of a crescendo of public alarm as the concern it provokes becomes self-generating, both through simple osmosis and by drawing to public attention more and more instances of the problem under discussion.[20] Eventually, however, the bubble deflates as the cost, and difficulties of action become apparent, and public interest moves elsewhere often leaving the original problem as unresolved, and the public as indifferent to it, as before the cycle began. We will see below to what extent environmental politics has followed this pattern.

2.4 The new environmentalism

The new environmental politics that came into being at this time varied in its details from country to country according to particular domestic concerns and political structures.[21] But an overall pattern is clear. With the partial exception of Japan, a strong environmentalist movement emerged everywhere. From the start the various national environmental movements established close links with one another and devoted significant attention to international issues and developments so that, at least at the level of its activists, the movement had a striking degree of international awareness and cohesion from the start.[22] An important and dynamic intellectual component was provided by scientists (many of them looking for a more humane application of their dis-

---

[19] See e.g. N. Cohn, *The Pursuit of the Millennium*, Secker and Warburg, London 1957. It is noteworthy that one well-informed commentator has observed that dark green environmentalism 'draws heavily upon scientific findings and theory, but its rationale is more religious than scientific' (L. Caldwell, *Between Two Worlds*, Cambridge University Press, Cambridge 1994).

[20] Downs, 'Up and Down with Ecology'.

[21] Kitschelt, 'Contemporary Social Movements'.

[22] Caldwell, *International Environmental Policy*.

cipline to public affairs than that offered by the Cold War) and in particular biologists, who provided many of the early gurus of the movement, notably Carson, Ehrlich, Commoner and Hardin. Ideologically and sociologically the environment movement had a great deal in common with the other 'new left' movements which sprang up at about the same time and with which, indeed, it shared a large common membership.[23] Those involved tended to be the young, the educated and the middle-class. The general thrust of environmentalism, in common with much of the rest of new left thinking, tended to be anti-establishment, anti-materialist, anti-growth and anti-technology. It is a caricature, but not an unrecognizable one, to describe its economic icons as organic farming, bicycles and windmills. The demand for tougher environmental regulation in fact stood in some contrast to its otherwise generally libertarian flavour.

Two of the biggest mass protests at the time, association with which has left an enduring imprint on environmentalist attitudes, were (in America) the anti-Vietnam War movement and (initially in Europe and later in America as well) opposition to nuclear weapons, which in the early 1970s extended to include opposition to nuclear power.[24] Indeed, the nuclear issue was in many ways the central formative experience of the new environmentalism. Here was a 'dread technology'[25] whose environmental risks seemed very apparent; whose chief proponents were those bêtes noires of the new left – big industry and big government; which was associated in the public mind with the nuclear bomb; and whose justification and explanation depended upon difficult probabilistic arguments hard to get over in public even when this was competently attempted, which often it was not. The upshot was a sequence of spectacular campaigns against nuclear developments across Europe and the US, many of them apparently successful, which have left opposition to nuclear energy a key tenet for most environmentalists ever since.

The environmental movement did not fit comfortably into the left-right spectrum of Western politics as it existed at the time. The right, with its traditional

---

[23] See Caute, '- *68 The Year of the Barricades*' for an excellent delineation of the distinction between the hierarchical, centralizing, class-based 'old left' and the youthful, libertarian, spontaneous 'new left'.
[24] McCormick, *Global Environmental Movement*.
[25] Kitschelt, 'Contemporary Social Movements'.

support for business and the military, probably had the greater difficulty in responding to the environmental agenda, even though those parts of it concerned with countryside and wildlife preservation were well-established 'conservative' causes. But the 'old left' too, including the trade unions, was very dubious about projects for environmental regulation which might, for example, have significant consequences for industrial employment. The British Labour Party accordingly dubbed the whole movement a 'middle-class preoccupation'.[26] The result was the emergence in a number of countries (most successfully in Germany) of 'green' parties campaigning on explicitly environmentalist programmes. In some countries, notably the UK and US, the growth of green parties was constrained both by a very well-established existing party system and a 'first past the post' electoral system which inhibited the election of representatives from single issue pressure groups. In those countries environmentalism made its political impact through pressure on existing parties to adapt their policy programmes to the new public mood.[27] In those countries too it tended to be the parties of the left, the Democrats in the USA and the Labour Party in the UK, which eventually took more of the environmentalist agenda on board.

> There was, and remains, a dichotomy between the real, but local, environmental concerns of the general public and the wider international vision of the environmental movement.

Japan was a special case. The citizens' movements that emerged there at the end of the 1960s remained intensely local and concentrated almost exclusively on local pollution problems.[28] But in most other countries too, and particularly the US, broad public concern about the environment (as opposed to views within the environmentalist movement itself) was principally driven

---

[26] P. Stone, *Did We Save the Earth at Stockholm?*, Earth Island Press, London 1973.
[27] McCormick, *Global Environmental Movement*.
[28] M. Schreurs, 'International Environmental Problems and Japanese Domestic Policy Making', paper presented at the International Environmental Institute Research Seminar Centre for Science and International Affairs, Kennedy School of Government, Harvard November 1992.

by anxiety about domestic pollution rather than the wider international issues.[29] There thus was, and remains, a dichotomy between the real, but local, environmental concerns of the general public and the wider international vision of the environmental movement.

The new environmentalism inevitably generated academic interest and activity. Scientists, economists, political scientists and sociologists all turned their attention to the new phenomenon. In particular it gave a new vogue to the age-old human endeavour to predict the future, this time with a generally gloomy twist. One of the best-known and most ambitious such exercises was that conducted by the Club of Rome. This group of some 70 eminent scientists, economists, businessmen and others used the most sophisticated computer modelling techniques available to look at mankind's prospects over the next century or so. Their conclusions, which sold 9 million copies in 29 languages, were stark. Exponential growth in rates of consumption of non-renewable resources would 'on the most optimistic assumptions' rapidly lead to their exhaustion. Population and pollution too would soar. Global calamity could be avoided only through very early and vigorous action to halt population growth and cut back on resource intensive industrial activity.[30]

At a more popular level a series of best selling books, notably *The Population Bomb* (1970) by Paul Ehrlich and *The Closing Circle* (1971) by Barry Commoner, offered gloomy prognostications on man's present and potential impact on the habitability of the planet. Ehrlich, in particular, asserted that 'the battle to feed all of humanity is over. In the 1970s the world will undergo famines, hundreds of millions of people are going to starve to death in spite of any crash programmes embarked on now.' It was against the background of predictions like these that U Thant, the Secretary General of the UN, made his much quoted statement in 1969:

> I do not wish to seem over-dramatic but I can only conclude from the information that is available to me as Secretary General that the members of the United Nations have perhaps ten years left in which to subordinate their ancient quarrels and launch a global partnership to curb the arms

---

[29] Springer and Constantini, 'Public Opinion and the Environment'.
[30] Donella Meadows, Dennis Meadows, J Randers and W Behurns III et al., *Limits to Growth*, New American Library, New York 1972.

race, to improve the human environment, to defuse the population explosion and to supply the required momentum to development efforts. If such a global partnership is not forged within the next decade then I very much fear that the problems I have mentioned will have reached such staggering proportions that they will be beyond our capacity to control.

2.5 Disagreements

The environmental movement was by no means unanimous in its thought. As tends to be the way with new and radical ideological movements, it was the cockpit of intense internal argument and factionalism. Perhaps the most important of these cleavages has been between what have come to be known as 'light green' and 'dark green' shades of environmentalism.[31] The less radical 'light green' position has been to underline the importance of our finding technical and regulatory solutions to the environmental problems caused by industrialization and economic development, without casting into question the desirability of industrialization and economic development as such. 'Dark greens' go much further, as illustrated by the contention of Porritt and Winners[32] that what is needed is 'a non-violent revolution to overturn our whole polluting, plundering and materialistic industrial society, and in its place to create a new economic and social order which will allow human beings to live in harmony with the planet'. Thus dark greens do not accept that technical solutions are enough. They argue, rather, that environmental destruction is intrinsic to the whole Western way of life, that the 'limits to growth' imposed by finite natural resources and the limited ability of the planet to absorb pollution will soon bring that way of life to an end, and that what is needed is a revolution in human values and aspirations which will replace contemporary materialism by recognition of the intrinsic value of the planetary ecosystem and a determination to live in harmony with it.

The importance of dark green thinking does not lie in the popular support it has enjoyed. It is quite clear that even at the peak of popular environmental

---

[31] A good account of this distinction is to be found in A. Dobson, *Green Political Thought*, Unwin Hyman, London 1990.
[32] J. Porritt and Winners, *The Coming of the Greens*, Fontana, London 1988.

enthusiasm, while Western publics wanted various types of pollution (especially local pollution) curbed, they were not disposed to abandon the pursuit of a higher material standard of living for that purpose. The role of dark green environmentalists, rather like that of the extreme left in relation to the wider movement for greater equality and social security at the start of the twentieth century, has been to act as a sort of 'revolutionary vanguard' in the wider environmental movement. Thus they have exercised ideological influence quite disproportionate to their numbers, partly through the force of their convictions and the uncompromising clarity of their ideology, and partly through the commitment and energy they bring to causes which attract much wider public support, such as the reduction of industrial pollution.

Quite apart from differences within the environmental movement itself, there were those outside who strongly questioned the imminence of the apocalypse. In 1972 John Maddox published *The Doomsday Syndrome*, which, as we have seen, sharply questioned the assumptions that underlay the fashionable predictions. He underlined the poor predictive record of Malthus and his followers and the recurrent surprises produced by human adaptability and ingenuity. In particular, commenting on Ehrlich's prediction of mass starvation Maddox noted that 'there is a good chance that the problems of the 1970s and 1980s will not be famine and starvation but, ironically, the problem of how best to dispose of food surpluses'.[33] Others, varying from Nobel Prize-winning scientists to the London *Daily Telegraph*, echoed his scepticism. The Club of Rome found its conclusions attacked in detail as unscientific and alarmist in a *Foreign Affairs* article entitled 'The Computer that Printed Out W*O*L*F*'.[34] Thus was launched a debate between those who feared the environmental worst in the absence of vigorous national and international action and those who inclined to the view that the problems were exaggerated, that normal economic and political feedbacks would deal with most of them and that intervention, even when justified, should be kept to an absolute minimum. This debate is, as we have already noted, still with us.

---

[33] J. Maddox, *The Doomsday Syndrome*, Macmillan London, 1972. Another much-cited presentation of similar ideas is Julian Simon's *The Ultimate Resource*, Princeton University Press, Princeton, 1981.

[34] Kaysen, 'The Computer that Printed Out W*O*L*F*', *Foreign Affairs*, Vol. 50, No. 4, 1972.

2.6 Government responses

Democratic politics tends to be better attuned to the public mood than to the intricacies of intellectual debate. By the end of the 1960s the intense public concern about domestic pollution issues was translating itself, at different speeds in different countries, into public action. Between 1970 and 1972, 14 environmental ministries or agencies were established in industrialized countries. In the US, a string of air and water protection measures in the 1960s culminated in the National Environmental Protection Act of 1969 which in principle made environmental quality a central concern of the federal government. By the early 1970s the US was spending nearly 2% of GNP on pollution control. In 1972, more than half the fifty states passed environmental legislation. There was a similar explosion of legislative activity in Japan, where in 1970 the 'Pollution Diet' passed 14 items of environmental legislation, and, at a slightly slower pace, in Western Europe.

The pressure for action on international issues was, of course, less. There were two reasons for this. First, as already noted, Western public opinion was principally exercised about domestic pollution problems, and that, consequently, is where the bulk of legislative action was concentrated. Second, there is no evidence that the political impact of popular environmental concern in the developed countries was paralleled to any significant degree in the developing countries or the communist bloc. In communist countries there are signs, as exemplified by letters to the press and support for conservationist organizations, of some public alarm about the environmental consequences of industrialization. But of course direct criticism of the party was not allowed and the concerns that were expressed seem to have made little, if any, impact on official policy.[35] In some developing countries too there were cases of localized popular concern about particular environmental problems. India, for example, saw protests in the 1960s against the pollution produced by a paper mill in Madhya Pradesh and a rayon factory in Maroor Kerala.[36] In Mexico, public concern about the deteriorating air quality in Mexico City in the 1960s led to a scientific symposium and comprehensive (looking) legisla-

---

[35] See eg. P.Pryde, *Environmental Management in the Soviet Union*, CUP, Cambridge 1991; and M.Feschbach and A.Friendly jr, *Ecocide in the USSR*, Basic Books, New York 1992.
[36] S. Khator, *Environmental Development and Politics in India*, University Press of America, Washington DC 1991.

tion.[37] There is also evidence that certain Third World elites, in particular scientific ones, with their greater links with the West and knowledge of intellectual trends there, were alive to the dangers of environmental degradation.[38] But there is no sign that these particular instances were more than isolated and unrepresentative special cases. They certainly did not point to the sort of generalized environmental alarm which, in the West, forced governments to give the issue the political priority it was receiving by the late 1960s. Except in certain rapidly industrializing localities, none of the factors identified above as fostering environmental concern in the West operated to anything like the same degree in the developing world. Indeed, even in the poorer parts of the West adherence to the new environmentalism appears to have been somewhat shaky. There is evidence, for example, that the Irish government remained keen throughout the 1960s to boost economic growth by attracting investment from certain types of polluting industry. An Irish politician of the time is quoted as observing that instead of 'the lads leaving Ireland for the big smoke...it'd be better for Ireland if they stayed here and we imported the smoke'.[39]

This does not mean that the international scene was completely somnolent. The intensity of public concern in the West prompted a significant shift of international rhetoric, and to a lesser extent spending, in an environmental direction. For example, environmental justifications were offered for the increasing proportions of US and other aid programmes devoted to population control projects.[40] Existing areas of international environmental business, notably the steady flow of expert negotiation on conservation issues and oceanic oil pollution, were also accelerated by increased public attention, producing in particular the 1971 Ramsar Convention on Conservation of Wetlands (which are an important habitat for wildfowl) and the second and third rounds of amendments to the 1954 oil pollution convention.

---

[37] C. duMars and S. Beltran del Rio, 'A Survey of the Air and Water Quality Laws of Mexico', *Natural Resources Journal*, 28, 1988.

[38] Caldwell, *International Environmental Policy*. See e.g. P. Pryde, *Environmental Management in the Soviet Union*, Cambridge University Press, Cambridge 1991.

[39] G. Leonard, *Pollution and the Struggle for the World Product*, Cambridge University Press, Cambridge 1988.

[40] S. Johnson, *World Population and the United Nations*, Cambridge University Press, Cambridge 1987.

What was to become a characteristic sequence of events also produced a cluster of new international environmental agreements at this time. In March 1967 the 118,000 ton oil tanker, *Torrey Canyon* ran aground just off the Cornish coast in south west England and deposited about 100,000 tonnes of crude oil into the sea in the largest oil spill the world had seen up to that date. The event, and the attempts at clean-up, received saturation news coverage, both in Britain and France (which were directly affected) and more widely. It also revealed large gaps in the international legal regime covering such spillages. There was no provision for compensation for the substantial financial damage done by the wreck, and it was not clear that under existing salvage law the governments concerned had any legal right to do anything to the wrecked tanker to try to limit the flow. The intensity of public interest throughout the developed world very swiftly caused governments to negotiate three treaties allocating the necessary rights to act in such cases, toughening liability and compensation arrangements, and creating an international fund to provide such compensation. In an interesting parallel private process, which seems to have been driven by tanker owners' fears of tougher obligations being imposed by coastal states if they were not seen to act themselves, the owners also established voluntary cooperative arrangements (beguilingly entitled TOVALOP and CRISTAL) to guarantee and fund compensation in the future.[41] Finally, as a further harbinger of the way things were going to go, the North Sea and north east Atlantic states, on the one hand, and the Scandinavian states, on the other, established regional treaties for cooperation in the event of an oil spill emergency. Thus was set the precedent for disaster-driven international environmental law, a pattern we will see frequently repeated.

> **Thus with the *Torrey Canyon* was set the precedent for disaster-driven international environmental law, a pattern to be frequently repeated.**

The growing developed-country expertise in negotiating marine pollution agreements had one other product at this time. A 1970 US government report

---

[41] E. Gold, *Handbook on Marine Pollution*, Assuranceforingen Gard, Rotterdam 1985.

on the environmental damage caused by the practice of dumping wastes at sea fed into a climate of Western opinion already alarmed by the *Torrey Canyon* disaster and other examples of marine pollution and rapidly produced first changes in US national law, and second a convention agreed among the states of the North Sea and north east Atlantic (the so-called 'Oslo Convention' of 1972) to limit and regulate such dumping in the future.[42] This convention set the precedent for much future international environmental legislation by splitting wastes into 'black' and 'grey' lists, dumping of the former being completely banned, and of the latter to be strictly regulated by parties to the convention. It also established a system of regular meetings of the parties to update the lists and other aspects of the convention in the light of developing technical knowledge. This, treaty, too is an early example of the 'level playing field' effect on environmental treaty-making. In this case a country (the US), having regulated its own industry in a certain way, then found itself under domestic pressure to extend these regulations internationally to offset the competitive disadvantage it had imposed – i.e. in the standard sporting cliché, to 'keep the playing field level'.

> **The crescendo of Western public concern required a more substantial and wide-ranging political response than was offered by specific agreements on oil pollution or wildlife conservation. A rather straightforward, low-cost and yet highly visible way to respond to such pressure is to convene an international conference.**

The crescendo of Western public concern about the environment as the 1960s approached their end, combined with the evident international dimension of the subject, required a more substantial and wide-ranging political response than was offered by specific agreements, however timely, on oil pollution or wildlife conservation. A rather straightforward, low-cost and yet highly visible way to respond to such pressure is to convene an international conference. Indeed, the idea of a major international conference on the environment

---

[42] US Council on Environmental Quality, *Ocean Dumping and National Policy*, Washington, DC 1970.

seems to have been circulating in the corridors of the UN by late 1967 and was formally raised there by Sweden at the end of that year. In the meantime, UNESCO, with a number of other UN bodies, had convened a high-profile meeting of experts on the biosphere in September 1968 which concluded, as expected, that the situation had reached 'a threshold of criticalness'. This paved the way for the UN General Assembly, on a proposal from Sweden, to agree at the end of 1968 to convene the UN Conference on the Human Environment. Sweden, which hoped to use the conference to spotlight its problem of 'acid rain' caused by industrial emissions elsewhere in Europe, also offered to host the conference, which was therefore to meet in Stockholm in 1972. Its formal aim, in the words of the 1968 General Assembly Resolution, was 'to provide a framework for comprehensive consideration within the UN of the problems of the human environment in order to focus the attention of governments and public opinion on the importance and urgency of this question'.

Chapter 3

Stockholm Conference

> *When you say you agree to a thing in principle*
> *you mean you have not the slightest intention*
> *of carrying it out in practice.*
> OTTO VON BISMARCK

There is a life cycle to UN conferences. Each one is born out of a political need to be seen to be doing something about a visible current problem. The announcement of the conference then generates high, if imprecise, public expectations. As preparations get under way it becomes clear to the negotiators, although not yet to the public at large, that the words which are, after all, all the conference will produce, will have to embrace a widely divergent range of national views, significantly diluting their eventual operational content. Current political differences which have little to do with the purposes of the conference will complicate, and may occasionally dominate, the negotiation. Governments will protest their attachment to the future welfare of humanity in public, while firmly (and, in their view, consistently) defending their national interests in private. As the preparatory period ticks away negotiators will start looking for some concrete decisions (normally institutional or financial) which can be used to meet the public appetite for concrete action. By this time ministerial speeches will be dwelling less on the unique importance of the conference and more on its role as part of a process. In a period of panic at the end compromise texts are cobbled together in which generalized language has often had to replace precise commitment, but in which large sums of money and eye-catching new institutions will, if possible, also figure. A grandiloquent package is presented to the waiting world and participants return to their capitals. It is left to a sceptical press to assess quite how much impact the whole affair has made on the problem it was originally designed to solve.

So it was with Stockholm, but with some extra complications.[1] Even at the time that it was announced it was clear that this was going to be an unprecedentedly large event by UN conference standards. The formal preparatory period was to last two years, from 1970 to 1972. All member countries were to be involved, as were all the component parts of the UN system. Each country was expected to prepare a comprehensive report on its environmental situation and the policies it was putting into place to deal with the problems. In the event 110 such reports were received. In addition, the conference secretariat commissioned a 'conceptual framework' for the conference. This emerged in print as *Only One Earth* by Barbara Ward and Rene Dubos.[2] It became a bestseller but, in line with its 'conceptual' mandate, is much longer on calls for collective international responsibility and 'loyalty to the earth' than on concrete ideas on how governments should deal with the environmental problems facing them. All in all there were over 100,000 pages of preparatory documentation, as well as 40 tonnes of documents at the conference itself. Meanwhile the continuing growth in the intensity of Western public concern about the environment, and particularly the events of 'Earth Day' in 1970, led Western countries in the General Assembly of that year to strengthen the formal aims of the conference from consciousness-raising (as set out in the 1968 General Assembly Resolution) to the more action-oriented purpose of serving 'as a practical means to encourage and to provide guidelines for action by Governments and international organizations'.[3]

3.1 The divergence

This intense Western enthusiasm was not shared elsewhere in the world. The communist bloc took the position that pollution was the product of capital-

---

[1] For the best accounts of Stockholm see W. Rowland, *The Plot to Save the World*, Clark Irwin and Co Toronto 1973; R.S. Berry, 'Only One World: An Awakening', *Bulletin of the Atomic Scientists*, September 1972; and P. Stone, *Did We Save the Earth at Stockholm?*, Earth Island Press, London 1973.
[2] Barbara Ward and Rene Dubos, *Only One Earth*, Peguin, Harmondsworth, 1972.
[3] UNGA Resolution No. 2581, 1969.

ism, and consequently a problem from which they did not suffer. The USSR and the East Europeans then dropped out of the preparatory process and the conference itself because of a separate Cold War dispute hinging on the international status of East Germany. The formal nature of this argument, and the fact that a way around the same difficulty had been found in other cases, strongly suggests that the USSR and its satellites did not see the Stockholm Conference, and the international environment more generally, as a high political priority at that time.

Stockholm was the first high-profile political attempt to draw the developing countries into international discussion of environmental issues. It thus put on display for the first time the central tension which has dogged global environmental discussion ever since. The major developing countries approached the conference with caution bordering on hostility. As noted above even in those developing countries where channels for the expression of public opinion existed there had been nothing like the upsurge of environmental concern that the West had seen. Indira Gandhi, for example, emphasized at Stockholm that poverty, not pollution, was the principal problem confronting India. As the Stockholm preparatory process began there were real doubts about how many developing countries would ultimately participate. Maurice Strong, the conference Secretary General, had to put the bulk of his efforts in the early months into persuading the developing countries to attend by assuring them that the conference agenda would extend beyond the environmental problems of industrialized countries to embrace developing country concerns too.

As part of this preparatory process the key developing countries rapidly evolved their own distinctive approach to the Stockholm agenda. This was most coherently summed up in the 'Founex Report',[4] which was produced by a group of developing country scientists and experts for the Stockholm conference, and was echoed, with remarkable solidarity, in political statements by developing countries in the run-up to, and through, Stockholm. The Founex Report was in many ways a seminal document, foreshadowing a debate which, 20 years later, is still vigorous. Its central argument can be summarized as follows.

---

[4] Miguel Ozorio de Almeida, Wilfred Beckerman, Ignacy Sachs, and Gamani Corea, *Environment and Development*, International Conciliation, Carnegie Document No. 586, New York, January 1972.

- Current environmental concern stems from the pollution and disruption of natural systems caused by high levels of industrialization.

- It was thus caused by, and principally affected, the industrialized countries.

- The developing countries also have a stake in these issues to the extent that they impinge on the global environment or on their own economic relations with the industrialized world, and to the extent that the developing countries themselves confront them in the course of their own economic development.

- However, the central environmental problems facing developing countries are different. They stem not from pollution but from poverty, disease, hunger and exposure to natural disasters.

- The solution to these problems was to be found through the process of economic development itself. In the rich countries industry might be the problem, in poor countries it was the solution.

> Both the Founex Report and subsequent statements made clear a profound underlying worry in the developing countries that Western concern about the environment could create pressures to slow industrial growth worldwide including in the developing world. In reaction they placed a heavy emphasis on sovereignty, and on the rights of countries to choose their own path of economic development

Both the report and subsequent statements made clear a profound underlying worry in the developing countries that Western concern about the environmental damage wrought by industrialization could create pressures to slow industrial growth worldwide including in the developing world. They were particularly concerned about the possibility of environmentally motivated restrictions on aid, investment or trade policies. In reaction they placed a heavy emphasis on sovereignty, and on the rights of countries to choose their own path of economic development, free of international interference for environmental or other reasons. These ideas thus loom large in the conference conclusions. The Brazilian ambassador to the US set out the developing country view well when he wrote in 1971:

Environmental deterioration, as it is currently understood in some developed countries, is a minor localised problem in the developing world. Nobody should expect to find an environment devastated by industrial activity where industries are so few and, more often than not, primitive. Evidently, no country wants any pollution at all. But each country must evolve its own development plans, exploit its own resources as it thinks suitable, and define its own environmental standards. The idea of having such priorities and standards imposed on individual countries or groups of countries, on either a multilateral or a bilateral basis, is very hard to accept.[5]

The Ivory Coast representative summed up the position more pithily in Stockholm by stating that his country would like more pollution problems provided they were evidence of industrialization.

All the signs are that this developmental twist given by the South to the environmental issue took the North by surprise. The Northern aim in calling for Stockholm had been to stage an international event focused on the problems of concern to their publics, such as marine pollution, overconsumption of natural resources and global population growth. At Southern insistence they now found the agenda broadened to include such issues as global poverty and aid levels. This fundamental divergence of approach meant that Stockholm, unexpectedly from the North's point of view, became a forum for the rehearsal of well-established North-South disagreements, garbed in green for this occasion, about the fairness of the international economic order and the need for greater flows of Western aid and technology. The developing countries held that if they were to achieve economic growth in a non-polluting way and make their contribution to protecting the global environment they would need extra funding from the West as well the modern technologies which would assist economic growth while keeping pollution down. This extra Western help would be justified by the need to compensate them for the costs of meeting higher environmental standards, and should be in addition to existing aid flows.[6] As the Founex Report gracefully put it, 'an emerging understand-

---

[5] G. de Araujo Castro, 'Environment and Development', in Kay and Skolnikoff eds, *World Eco-crisis*, Wisconsin University, 1972.
[6] Rowland, *The Plot to Save the World*, esp. chs 3 and 4.

ing of the indivisibility of the earth's natural systems on the part of the rich nations could help strengthen the vision of the human family and even encourage an increase in aid to poor nations' efforts to improve and protect their part of the global household.'

This was the first (and, by later standards, fairly muted) appearance of the argument that environmental action by the developing countries must depend on more help from the West. The Western reaction was not forthcoming. The major donors, notably the US, maintained firm hostility to these 'compensation' and 'additionality' principles, pointing out, in particular, that the compensation principle was difficult to square with a key tenet of the emerging environmental thinking, the 'polluter pays' principle. As is usual with arguments about money this disagreement was not easy to resolve. It ran right through the preparatory process and ended only when the US was voted down. The compensation and additionality principles, and encouragement for technology transfer, thus figure in the concluding documents of the conference, but in a non-binding way.[7] There is no evidence that they have had any impact on subsequent flows of Western aid and technology.

> The 'compensation' and 'additionality' principles, and encouragement for technology transfer, thus figure in the concluding documents of the conference, but in a highly non-binding way. There is no evidence that they have had any impact on subsequent flows of Western aid and technology.

Another potential area of disagreement between North and South was population policy. To some extent tension over this topic was defused by the standard diplomatic technique of deferring consideration of it to a later conference (the World Population Conference of 1974). Nevertheless, concern about rapid population growth (which was largely taking place in developing countries) was one of the principal themes of the Western environmental movement and was, therefore, echoed in Western official positions in the run-up to Stockholm. This provoked a reaction among developing countries who, it emerged, fell into two broad streams of opinion on the issue. There were those (notably

---

[7] Cf. Stockholm Conference Conclusions, Principle 11, and Recommendations 103(b), 107, 108 and 109.

certain Latin American and Muslim countries) who resisted the idea that growing population was necessarily a source of environmental degradation. They were, however, outnumbered by those (including in particular China and India) that had already accepted the desirability of population policies, and in a few cases were already pursuing them (India's first birth control programme began in 1951), but firmly denied the West the right to dictate such policies to them. Indeed, in their view the West's obsession with population policy was little more than an attempt to distract attention from the real source of most environmental degradation – the West's own high consumption levels. Thus the Chinese delegation emphasized that the growth of China's population had not prevented an even faster growth in living standards there, while Indira Gandhi pointed out with force that one inhabitant in an affluent country caused much more pollution and used many more resources than one inhabitant in a poor country.[8] The Stokholm Conference agreed by a substantial majority to recommend intensified, but unspecific, UN activity on population control.

3.2 Other issues

While these differences between North and South dominated the preparatory process, they did not do so to the total exclusion of other issues. The Swedes pressed, as expected, for strong conclusions on acid rain but met resistance (on the grounds of lack of scientific evidence) from those countries from which the pollution was supposed to be coming, and had to content themselves with a recommendation that the problem be monitored. Similarly, Scandinavian efforts to brand supersonic air transport with damaging the ozone layer were implacably blocked by the UK and France. Brazil, using arguments based on both national sovereignty and third world solidarity, diluted to meaninglessness proposed strong conclusions on deforestation.[9] There was a long-running dispute about atmospheric testing of nuclear weapons which pitted the US and New Zealand on the one hand against China and France on the other. The debate about marine protection was largely con-

---

[8] Indeed, P. Ehrlich *The Population Explosion*, Ballantine Books, New York, 1990, has calculated that the environmental impact of one American is the same as that of two Swedes, three Italians, 13 Brazilians, 35 Indians, or 280 Chadians or Haitians.
[9] R. Clarke and L. Timberlake, *Stockholm Plus Ten*, Earthscan, London 1982.

ducted between those, notably Canada, who wished to see coastal states given greater authority over activities in their neighbouring waters, and states with substantial marine interests, such as the UK and the US, who wished as far as possible to maintain the freedom of the seas. This agenda item was given an extra twist by Brazil and Mexico, who tried to delay agreement on the negotiation of a marine dumping convention until they got support for their wish to extend their coastal waters to 200 miles. There was a farcical debate about whales. The US delegation, largely to please the US press and NGOs, launched and had adopted (to cheers from the public gallery), a demand for a ten-year moratorium on whaling. Within a month, however, this proposal was quietly killed by the International Whaling Commission (the body which, as everyone knew, was formally responsible for the regulation of whaling) with a number of countries reversing in private the support for the proposal they had given in front of the TV cameras in Stockholm. There were vitriolic speeches by African countries on apartheid and by Arab countries on the plight of the Palestinians; and there were eloquent attacks by Sweden and China on US policy in Vietnam (made environmentally relevant by focusing on the use of defoliants) followed by equally eloquent US rebuttals.

3.3 Form of the conference

This intoxicating mixture of business as usual with the radically new subject matter of the environment extended to the shape of the conference. In form it was ostensibly a traditional UN conference, albeit a large one. One hundred and fourteen nations participated, sending about 1,200 delegates. Most of these were at ministerial or equivalent level, and there were two heads of government present: Olaf Palme from the host country and, interestingly, Indira Gandhi from India.

This conference, in striking contrast to most UN meetings, seems to have generated a feeling of excitement and anticipation. This was the first UN theme conference. Many of the participants were young, and as yet uncynical about the tortuous and wearisome ways of international negotiation. Many hoped that the conference would mark a breakthrough on the environment and a major step towards effective international action to succour the biosphere. Stockholm had two other features which were to become highly characteristic of international environmental business. The first was the extent of media interest. There were over 1,000 journalists present, and the substantial coverage which they produced further raised the public profile of environmental issues, both in the West and more widely. It also directly affected the formal business of the conference, as has been noted above in the case of whaling. Other issues (for examples, the arms trade) produced similar 'green grandstanding'.

The second distinctive feature of Stockholm was the involvement of non-governmental organizations. The rapid growth of the environmental NGO movement in the West, and its international character from an early date, has already been mentioned. The movement was such an evident force in Western environmental policy-making that the conference organizers decided to give it a major role at Stockholm, and throughout the preparatory process. They therefore organized an 'environmental forum' for non-governmental debate and activity to run in parallel with the conference proper. Some 500 NGOs participated. These included all the mainstream Western environmental organizations as well as a huge variety of scientific groups and other special interest groups concerned with the environment. There were also a few NGOs from developing countries. This mass of bodies pursued debate in their own forum, which displayed an energy and enthusiasm often depressingly absent from the formal negotiations, but also took on a heavily new left and third-worldist flavour (so intense that, for example, even so highly esteemed a figure as Paul Ehrlich was shouted down because his concerns about over-population were seen as being anti-third world). By all accounts the forum was a colourful and stimulating experience for the participants, offering political demonstrations, endless debate, and nearly being closed down early because it couldn't pay its bills.

The NGOs were also permitted to make a formal statement to the conference. This document offered a tough (but, as seen from 1993, by no means absurd) overview of global environmental problems. It included a carefully balanced sentence on population, but also emphasized 'the overriding necessity of moving at once to a significant redistribution of the world's resources in favour of the developing countries'. More importantly, the NGOs were able to follow the conference discussions and, through their access to delegations, feed in their views. In particular they made the highly successful innovation, which they have followed at every major environment conference since, of publishing a conference newspaper, *ECO*, which became required reading among the delegates and thus exercised some real influence on the proceedings (as, for example, in the run-up to the whaling debate).

The net effect of these two factors at Stockholm, close media attention and intimate NGO involvement was to make the negotiation of international texts a much more open process than it had been before, both in terms of visibility and in terms of exposure to public pressure. Moreover, that pressure, as expressed by NGOs and by Western public opinion, operated almost exclusively in favour of greater environmental concern and action. It is difficult to believe that this did not push Western governments further in the direction of incorporating environmentalism in the Stockholm texts they might have been willing to go in a more conventional, and private, negotiating process.

3.4 The agreed texts

The Stockholm process was designed to produce what we have defined as a level one response to world environmental problems – political statements intended as a basis for later, legally binding, action at levels two and three. The key products of the conference were three in number:

- the 'Declaration of the United Nations Conference on the Human Environment' (the Stockholm Declaration); consisting of 26 principles intended as a foundation for future development;

- the 'Action Plan for the Human Environment', consisting of 109 recommendations for governmental and intergovernmental action across the full range of environmental policy ranging through species conservation, forests and atmospheric and marine pollution to development policy, technology transfer and the impact of the environment on trade;

- resolutions agreed by the conference to set up a new UN environment body and fund.

The first of these items, the Stockholm Declaration, is very visibly the child of the North-South tensions described above. Its references to the need to protect resources and limit pollution are carefully balanced by references to the necessity of economic development and the sovereign right of states to exploit their own resources. Principle 23 underlines that standards set by industrialized countries may not be appropriate for developing countries, and principles 9-12 assert the need of developing countries for Western aid, technology and other help. Population policy must only be as 'deemed appropriate by governments concerned'. States must strive to reach agreement on the elimination of nuclear weapons. Overall one would have expected a text of such high generality, and therefore (as we have already noted) little concrete bite, to have been easily negotiated. This was not so. Since it was intended (at least by some) as a reference for future international law-making, innocuous-looking phrases took on apocalyptic importance. The delicate North-South balance in the text was not easy to find in the course of the preparatory process, and once found, was passed 'like a piece of Dresden china' to the Stockholm Conference with an earnest hint that discussion not be reopened.[10] It nevertheless was reopened and provoked an epic debate, much of it between China and the US, which ended in the closing minutes of the conference and whose overall result was to introduce extra language underlining the special needs of the developing countries, as well as references to a range of other political issues such as apartheid and colonial oppression.

While most of the Stockholm text is so vague as to be useless for legal purposes, two points discussed in connection with it were not. Indeed, if unambiguously introduced into international law they would have very extensive consequences. The first was to have been principle 20. This would have required states to supply information on any of their internal activities to neighbouring states which might be environmentally affected by those activities. This principle was demanded by Argentina (with widespread support) partly because of its alarm about a Brazilian dam project on the Pirana river. Brazil firmly opposed it. This is the one item which could not be settled at Stockholm but was remitted to the subsequent General Assembly where, faced with total Brazilian opposition, the text was dropped.

The second item did get into the final text as Principle 21. This has two components which on the face of it look difficult to reconcile: the sovereign right of states to exploit their own resources in line with their own environment policies, and their responsibility to ensure that activities in their control do not cause damage to the environment of other states. Unlike the rest of the Stockholm declaration, this principle has been much quoted since, and incor-

---

[10] Rowland, *The Plot to Save the World*.

porated in the preambular language of a number of environmental treaties.[11] The evidence, however, is that this has happened because the principle articulates the tension between the demand for economic development and concern about consequential transboundary environmental damage, without offering any concrete indication to how this tension should be resolved in particular cases. Developing countries when citing it subsequently have tended to dwell on the 'sovereign right' aspect, while developed countries have focused on the 'avoidance of transboundary damage' aspect. In practice, therefore, its concrete effect has been very limited.

The second product of Stockholm, the Action Plan, is long and looks comprehensive. It runs to almost 40 pages and includes recommendations on the whole gamut of environmental concerns ranging from species destruction to marine pollution and from forests to whales. There is no evidence, however, that its impact on subsequent events has been proportionate to its bulk. Such texts tend to become a Christmas tree on to which every country endeavours to hang its own pet projects. There is no doubt that many of the ideas in the text, particularly those for extra environmental research and monitoring, were well justified and have subsequently been pursued, at least to some extent. But these ideas exist only in the form of recommendations and it is finally a matter for individual states and multilateral organizations to decide how far, if at all, they wish to implement them. On the 'concrete commitments' test (noted in Chapter) this text scores very low. In particular, as already noted, there is no evidence that the hard-fought language on aid has had any impact on actual aid flows. It is questionable, too, how

> The second product of Stockholm, the Action Plan, runs to almost 40 pages and includes recommendations on the whole gamut of environmental concerns. There is no evidence however that its impact on subsequent events has been proportionate to its bulk. Such texts tend to become a Christmas tree into which every country endeavours to hang its own pet projects. On the 'concrete commitments' test this text scores very low.

---

[11] E.g. the London Dumping Convention and the Biodiversity Convention.

much effect the rest of the text has had on the subsequent development of national and international policy. It does contain a number of very concrete recommendations for environmental treaties (as we will see below) but these all turn out to be encouragements for items of business which were already under way or planned.

3.5 The institutional outcome: UNEP

Finally, there are the resolutions agreed at the end of the conference. These marked the end of a long and difficult debate. One of the aims of Stockholm was of course to set guidelines for future handling of environmental issues by the international system. A number of UN agencies already had environmental responsibilities of one sort or another: the IAEA for atomic energy, FAO for agriculture and forests, UNESCO for science and WHO for environmental health. In the manner of such organizations they tended to see the upsurge of Western political interest in the environment as a justification for expanded activity in their particular areas and, they hoped, for larger budgets. Other multilateral organizations, notably the big development funds, found their activities being examined from a point of view which was deeply dubious about the developmental philosophy which they had hitherto been pursuing. The World Bank, for example, experienced sharp environmental criticism of a number of major dam and other projects; in response, it had appointed its first environmental adviser in 1970 and announced (at Stockholm) that henceforth environmental assessment would be an integral part of project preparation.

Thus by about 1970 most of the major multilateral institutions were adjusting their policies in one way or another to the new demand for environmentalism (it is, incidentally, a rather clear demonstration of the political predominance of the West in these institutions at that time, largely for budgetary reasons, that this shift in policy took place despite the often explicit doubts expressed by developing countries over, for example, the 'greening' of aid.)[12] But in the Western public and political view this was by no means enough. While individual UN agencies might cover particular aspects of the environment, their various efforts were (as befits bodies which do not communicate as well as they might with one another) ill-coordinated and often overlapped or com-

---

[12] de Araujo Castro, 'Environment and Development'.

peted. The absence from the UN system of any central focus for the environmental activities of the agencies, or any means to coordinate their activities, became more and more apparent. Moreover, there were strong arguments for an international fund with specifically environmental objectives to sit alongside the already existing developmental funds. Significant Western public and political pressure for a new UN environmental institution built up rapidly.[13]

Institution-building within the UN system is not an easy process. There are too many rival interests – Who will pay? Where will it be sited? What nationalities will get the top jobs? How will its responsibilities fit in with those of existing agencies? The most immediate concrete product of Stockholm was an end to what could have been an interminable debate about a future UN environmental institution. Proposals initially on the table ranged from a new UN 'Environment Council' through a new specialized agency (or, as U Thant suggested, an environment 'super agency') to no more than a unit to allocate responsibilities between the existing agencies. The developing countries were broadly hostile to anything which would significantly impede the UN system's developmental activity. The existing agencies in the field did not wish to see a new player on their turf, certainly not one of full UN agency status, let alone some sort of overseeing council.[14] The major developed countries were torn between the need to produce something big enough to make the necessary political splash at Stockholm while also keeping the budgetary costs down.

Maurice Strong, Secretary General of the Stockholm Conference, managed to find a way through these conflicting forces partly as a result of intense secretariat preparation of the institutional outcome of the conference. In particular, a series of expert working groups drew together the key concepts of the eventual solution to the institutional conundrum: the UN Environment Programme (UNEP) and the Environment Fund it administers. In order to meet the concerns of the other agencies this was not set up as a full agency but as a unit within the existing UN structure. In order to meet the concerns

---

[13] See e.g. G. Kennan, 'To Prevent a World Wasteland, a Proposal', *Foreign Affairs*, Vol. 48 No. 3, April 1970.

[14] This was particularly true of the IAEA, which clearly saw environmentalist concern about nuclear issues as a threat to its own autonomy (cf. Stone, *Did We Save the Earth at Stockholm?*).

of the developing countries it was given a large governing council (to permit wide developing country representation) and, after an argument, sited in Nairobi. The developed countries could nevertheless point to their achievement in getting agreement to a new environmental body. Its principal function was to play a 'catalytic and coordinating' role and so act as a focal point for environmental activity in the UN system, notably for the various research and monitoring activities called for in the Stockholm conclusions. Its head would chair a new UN environment coordination board whose aim was to achieve better coherence among the activities of the other UN agencies. Those concerned to keep the costs down were placated by making UNEP small (its organizational budget until the end of the 1980s was less than $5 million per year) and by making contributions to the associated environment fund voluntary. Nevertheless, it did more or less achieve its objective of raising $20 million per year for the first five years.

3.6 Assessment

In assessing the overall consequences of Stockholm it is helpful to distinguish between the formal and the informal products of the conference. With regard to the formal products, and with the partial exception of UNEP, it is difficult to argue that they have had more than a marginal effect on the subsequent history of international environmental action. The Stockholm Declaration has not become the guiding hand for international law that it was intended to be. We have seen that even the much-quoted Principle 21 has tended to fuel debate rather than resolve it. The action plan provided material for the early work plans of UNEP and other relevant UN agencies, and references to it appeared for a short time in a variety of national and international fora. But it is very hard to argue that it did more than catalogue existing environmental concerns and activities, rather than redirect them or push them forward in any significant way. When UNEP assessed progress under the action plan ten years later the conclusion was that it had 'only been partially implemented and the results cannot be considered as satisfactory'. Given the traditionally rose-tinted phraseology of UN texts, this amounts to a profound admission of failure. There had been some progress in the easy and relatively cheap areas – monitoring, information exchange, research and consciousness-raising –

but not in those areas which cut across significant economic or developmental interests or required major changes of policy or administration. These, then, were level one responses to international environmental problems which have achieved little follow-up at level two or level three.

Even UNEP itself was initially no more than a minnow in the piranha-filled UN pond. In its early years its principal roles were consciousness-raising and collaboration with the programmes of other agencies and national governments. Its influence within the UN system has suffered as a result of its (in UN terms) peripheral geographical location. The sums available to it (and money is of course a key source of influence for any UN body) have been negligible by comparison with those available to other relevant bodies. The FAO and UNESCO, for example, both have annual budgets of over $200 million, and UNDP programme expenditure stands at over $800 million annually. UNEP pretensions to coordinate the environmental activities of the larger agencies were rapidly shipwrecked on the fierce independence of those agencies and its own small status within the system. The Environment Coordination Board, through which it was supposed to exercise this role, was wound up in 1977.

Thus, in terms of its formal product it is difficult to view Stockholm as much more than a cosmetic event. It demonstrated to Western publics that their governments were taking the international environment seriously. It exposed developed countries and developing countries to the gulf between their respective views of the environmental issue and demonstrated the impossibility of conducting global discussions on environmental problems without also facing the developmental issues linked to them. This lack of fundamental agreement meant that many of the conclusions were vacuous or doomed to non-implementation. The institutional and financial outcomes, while eye-catching, were marginal. Over subsequent years international activity on the environment would depend far more on political pressures in individual states than on any agreement or machinery that Stockholm created.

On the other hand, Stockholm was the first full-scale display of the new environmental diplomacy, conducted largely in the open with intense media and NGO attention. The mere fact of the conference also forced many governments to focus seriously on environmental issues in a way that they had not done before. This had consequences which we will look at below.

Chapter 4

From Stockholm to Rio: Domestic Developments

> *People are fickle by nature: it is simple to convince them of something but difficult to hold them in that conviction.*
> NICOLÒ MACHIAVELLI

4.1 Global trends

Environmental history is often presented as a steadily rising tide of pollution and resource depletion accompanied by a growing state of public concern. Neither half of this picture is accurate. By the time of Stockholm some forms of environmental damage, at least in the West, were under active repair; and the conference marked a high point in Western public attention to the environment, whereafter there was a swift relapse.

This was not because mankind's pressure on natural resources diminished. On the contrary, the impact of human activity continued to grow steadily.[1] In the period 1970-90 global population rose from 3.7 billion to about 5.3 billion. Energy consumption grew even faster from the equivalent of about 5 billion tonnes of oil per annum in 1970 to about 8 billion in the early 1990s. Nuclear production of electricity rose twentyfold. The number of motor vehicles more than doubled to 540 million. Gross world product (in constant 1990 dollars) rose from $8 trillion to $19 trillion. Mankind's activities came to constitute an increasingly significant proportion of the great natural cycles

---

[1] The figures in what follows are drawn from M. Tolba et al., *The World Environment*; *World Resources 1992-93*; and J. Ausubel and D. Victor, 'The Environment since 1970', Paper circulated to the Harvard International Environment Institution Research Seminar 1992.

of the earth. By the early 1990s people were consuming or pre-empting (for example, by concreting over fertile soil) about 40% of the entire planetary 'natural product' from the photosynthesis of plants.[2] Human activity now fixes (largely because of artificial fertilizers) almost as much nitrogen as does nature. Humans put about as much sulphur dioxide and nitrogen oxides into the atmosphere as all natural sources combined; and the proportion of carbon dioxide in the atmosphere has increased by about a quarter over the last century as a result of anthropogenic emissions.

Yet despite all of this economic growth, the gloomy predictions of the Club of Rome and other neo-Malthusians that mankind was rapidly exhausting the resources of the planet did not come to pass. The 'green revolution' brought about by the breeding of more productive and disease-resistant crop varieties, and 75% increase in the worldwide use of chemical fertilizers, have (as we will see in more detail below) enabled food production to keep up with population growth even in the poorest parts of the world. As predicted by Maddox, the principal agricultural problem facing the richer regions – notably the US and Western Europe – has become overproduction. The Club of Rome identified seven minerals, fast-disappearing reserves of which, they argued, were likely to impose an early brake on human economic progress. Proven economic reserves of all of those minerals now, after 20 years of accelerating consumption, are higher than when they wrote.[3] In the case of oil, for example, proven reserves have risen from 550 billion barrels in 1970 to 900 billion at present, even though 600 billion barrels have been consumed in the meantime. The World Bank now estimates that global fossil fuel resources (especially coal) are sufficient to supply the world, even taking expected economic growth into account, for the next century at least.[4] This is not to say that supplies will never run out, simply that those who argued in the 1970s that rising prices would produce a wider search for reserves, more efficient technology and, eventually, a shift to substitutes have been vindicated by events. Even in the case of forests, another resource whose destruc-

---

[2] P. Vitousek et al., 'Human Appropriation of the Products of Biosynthesis', *Bioscience 36*, 6 June 1986: 368.

[3] W. Beckerman, *Economic Development and the Environment, Conflict or Complementarity*, World Bank, Washington DC 1992.

[4] World Bank, *World Development Report 1992*, Oxford University Press, Oxford 1992.

domestic developments 53

tion has been widely advertised, the evidence is that the overall amount of tree cover on the planet has remained roughly constant over the past 20 years, with the destruction of tropical forests being roughly offset by new growth of temperate forests in the richer North.[5]

The sustained economic growth of the past 20 years has inevitably had a dramatic effect upon the environment and environmental politics. The precise consequences have varied from country to country and region to region, but it is possible to identify some general features. We focus below on the impacts within individual state borders, leaving international issues to the next chapter.

4.2 Developments in the OECD

In the rich North, where intense public concern led to a spate of environmental law-making in the late 1960s and early 1970s, it is clear that those new laws have begun to reverse certain types of environmental damage over the ensuing two decades.[6] Environmental spending in all developed countries has risen to somewhere between 1% and 2% of GNP, about evenly divided between the public and private sectors. As a result access to clean water, adequate sanitation and municipal waste disposal in developed countries is now more or less universal. Air quality has substantially improved. Particulate emissions have fallen by 60% and sulphur dioxide emissions by over a third (and in Japan by over 80%). Lead emissions are down by 85% in North America and 50% in most European cities. Other pollutants such as DDT, polychlorinated biphenyls and mercury are also down significantly. The quality of inland rivers and lakes is marginally up and in some high-profile cases of high pollution (the Rhine, the Mersey, the Great Lakes) there have been sharp improvements. Forested areas and national parks have expanded in almost all countries. Concern for the environment has also contributed to some significant changes in lifestyle and in industrial structure. The US environmental protection industry now has an estimated annual turnover of $120 billion.

---

[5] *World Resources 1992–93*. This shift does not of course compensate for the loss of the much richer biological diversity in the tropical forests.
[6] Figures that follow are drawn from OECD *The State of the Environment 1991*, OECD, Paris, 1991, though it should be noted that spending figures in particular are very uncertain in the absence of agreed definitions of 'environmental spending'.

The US now recycles three times as much glass, twice as much paper and 50% more aluminium than it did two decades ago, and similar trends are apparent in Europe and Japan.[7]

There is no doubt that some of these improvements are the result of economic restructuring, and in particular of the shift of significant amounts of heavy and polluting industry away from the West to the developing world. The evidence suggests, moreover, that a proportion of this shift is the result precisely of tougher environmental regulation in the West[8]. They are nevertheless an impressive demonstration of the ability of Western polities to respond to environmental problems at the national level when motivated to do so.

An important recent feature of the way Western countries tackle these problems has been a gradual shift from simply setting emissions limits for pollutants to more market oriented techniques which usually have the advantage of achieving the same environmental objective in a more flexible and economically efficient manner. Thus, in what was probably the first conscious application of this approach, the UK and seven other OECD countries have now all brought down lead emissions by imposing a higher tax on leaded than an unleaded petrol. Germany imposes fees on pollutants in industrial wastewater. France and Japan have 'taxes' on air pollution. Sweden offers tax incentives for the purchase of low-pollution cars. The US has a system of 'emissions trading' for power plants. Overall this style of environmental management has mushroomed, with 153 different economic instruments reported to be in place in the OECD countries in 1988 (although economists agree that there still remains enormous scope for its further application).[9] The 1990s have even seen the proposed extension of this kind of approach to the regional level with the European Community proposal for an energy/carbon tax to cut CO_2 emissions within the Community.

---

[7] UNEP, *'United Nations Environment Programme Environmental Data report 1991*, Basil Blackwell, Oxford.

[8] R. Lucas D Wheeler and H Hettiage et al., 'Economic Development, Environmental Regulation and the International Migration of Toxic Industrial Pollution 1960-1988', in P. Low, ed., *International Trade and the Environment*, World Bank, Washington DC 1992.

[9] Tolba et al., *The World Environment*; T. Tietenberg, 'Economic Instruments for Environmental Regulation', in D. Helm, ed., *Economic Policy towards the Environment*, Blackwell, Oxford 1991.

However, not all the environmental challenges facing the West, even at the national level, are yet under control. Emissions of nitrogen oxide, which, with sulphur oxides, are a principal cause of acid rain, have risen by 12% since 1970 in the OECD as a whole (except Japan) largely as a result of increasing use of road transportation. Although the area of Europe subject to acid rain is now falling, the acidity of European soils continues to rise and certain areas of Europe and North America continue to receive up to ten times natural acid precipitation. OECD municipal wastes grew by 28% between 1975 and 1990, posing growing problems of disposal. Groundwater and enclosed seas are being polluted, notably by fertilizers, pesticides, sewage sludge and heavy metals from industrial discharges. Action to improve air quality in major urban areas such as Tokyo and Los Angeles has been counterbalanced by growth of population and car use.[10]

The intensity of Western public concern about environmental issues, having rocketed in the late 1960s, fell back in the 1970s with a speed which, while less dramatic than its rise, still seems excessive when compared to the rate with which the problems were being dealt with. Quite a lot of this shift of attention no doubt reflected the emergence of other issues which by their very nature weakened the public appeal of environmentalist arguments, notably the oil shocks and economic slowdown of the early 1970s. It is therefore noteworthy (and an important variation from the 'issue/attention cycle' described in Chapter 2) that public concern did not revert to the low base from which it had started. Throughout the West the environment was now a familiar issue in a way it certainly had not been a decade earlier, although the period saw a noticeable shift in emphasis from the supposed threat of global resource depletion to the much more evident and more demonstrable problems of pollution and degradation of the global commons. In general, the NGOs maintained the membership they had built up in the 1960s, but their vertiginous rates of growth fell off, and in some cases turned negative. Media coverage, too, fell back, although a steady stream of accidents brought the issue back to the front page and the television screen at regular intervals: the dioxin leak at Seveso in Italy in 1976, the Amoco Cadiz oil spill in 1978, the Three Mile Island nuclear accident in 1979, the catastrophic leak of methyl

---

[10] Ausubel and Victor, 'The Environment since 1970'.

isocyanate in Bhopal in India in 1984, the Chernobyl nuclear explosion followed by the Basle fire and pesticide leakage into the Rhine in 1986, and the environmental depredations of Saddam Hussein during the Gulf War of 1991.

Nevertheless, in the US the proportion of opinion poll respondents suggesting 'the environment' when asked which were the most important problems facing the country fell from about 40% in 1970 to about 10% by 1974-5 and drifted even lower over the rest of the decade.[11] In Japan the number of environmentally concerned citizens' movements fell almost as dramatically as it had risen.[12] In the UK and France public interest dipped similarly. In New Zealand support for the world's first green party fell from 4% in 1975 to less than 1% in 1981. Only in northern continental Europe did the level of popular environmental concern stay anywhere near the point it had reached in 1970, and indeed voting support for 'green' parties in national elections rose steadily over the late 1970s and 1980s in Belgium, West Germany and Switzerland.[13]

It has been plausibly suggested that this movement of Western public opinion reflected the fact that by the early 1970s the environment had become a consensual social issue like concern for human rights or famine in Africa.[14] At the level of abstract principle there was little internal debate about the desirability of getting pollution down or conserving natural resources. The government could be expected to get on with it and so there was less reason for the general public to remain actively concerned. It was an issue which could uncontroversially be the subject of school projects, church sermons and royal speeches.

This does not mean that there was no debate about how to tackle particular environmental problems, and who should bear the costs. Throughout the 1970s Western countries saw a string of battles on such issues, in general between environmental groups and industry. A characteristic and long-running example was the American argument about fuel efficiency in motor vehicles, in which the US automobile industry, while acknowledging the desirability of

---

[11] R. Dunlap, 'Trends in Public Opinion towards Environmental Issues 1965–1990'. *Society and Natural Resources 4* (1991) 285–312.
[12] Krauss and Simcock, 'Citizens' Movements'.
[13] Kitschelt, 'Explaining Contemporary Social Movements'.
[14] Dunlap, 'Trends in Public Opinion towards Environmental Issues 1965–1990'.

more efficient cars, staunchly resisted as impractical and ruinously expensive the standards proposed to achieve this end. This became the characteristic stance of industry in these debates: to acknowledge the need for environmental protection, in general terms, while arguing against the particular measures proposed to achieve it, often on grounds of cost, superfluity or scientific inadequacy (or all three) – reinforcing the conclusion that by the early 1970s the environment had become a 'motherhood' value which even those who had most to lose from rising environmental standards could not afford to attack frontally. In each country the same arguments were fought out in characteristic national ways. The issue of workplace limits for vinyl chloride (a carcinogenic industrial gas), for example, produced a major public confrontation between industry and the authorities in the US, while in Japan it was settled privately and quietly at meetings between representatives of industry and the government.[15]

In the case of the US, the existence of a social consensus on the need for environmental protection is rather confirmed by events after 1980, when the Reagan administration came to power and announced its intention of loosening environmental standards in the interests of faster economic growth. This provoked a rise in public concern for the environment, as expressed in US opinion polls, as well as resumed growth in environmental groups. Membership of the Sierra Club, for example, more or less doubled between 1980 and 1985.

4.3 The developing world

The nature of the problems

In the developing world, even after all the caveats have been made about the huge variety of national situations and experiences, the broad picture was radically different. In looking at this picture[16] it is helpful to divide problems into two broad groups: those stemming from poverty and population growth,

---

[15] S. Pharr and J. Badavacco, 'Coping with Crisis: Environmental Regulation', in T. McCraw, ed., *America vs Japan*, Harvard Business School Press, Harvard.
[16] Data drawn from the *World Development Report 1992*, World Bank, Washington DC, 1992 and *World Resources Guide to the Global Environment 1992/3*, Oxford University Press, Oxford, 1992.

and those resulting from economic growth and industrialization. Evidently, as countries – for example in Asia and Latin America – have moved up the ladder of economic growth, the difficulties they face have begun to shift from one type to the other.

Population growth by itself is not an environmental problem. It becomes a major environmental factor, however, when it begins to place excessive strain on the resource base which must support the extra people.[17] As human numbers have grown such strains have become increasingly apparent in much of the developing world, notably in the form of rising pressures on the two crucial resources of clean water and cultivable land. Even though huge efforts have been put into the provision of uncontaminated water (which was supplied to 1.6 billion more people from the early 1980s to the early 1990s) access has in fact barely kept pace with population growth and remains strikingly low in large areas of sub-Saharan Africa, south Asia and Indonesia. The quality of river water in low-income countries (as classified by the World Bank) has tangibly fallen in the past decade, while in medium-income countries it has remained constant (and low) as clean-up measures have been offset by increased use. The effect is that waterborne diseases remain rife, with over 1 billion cases annually and 3 million infant deaths from diarrhoea, roundworm and related sicknesses.

As far as cultivable land goes, it has been estimated that some 1.2 billion hectares – 11% of the earth's vegetated surface – has suffered soil degradation, since the 1950s, mostly as a result of land clearance, overgrazing and other environmentally improvident agricultural practices.[18] This damage, which is mostly concentrated in Africa and Asia, has brought about falls in fertility and agricultural productivity estimated at somewhere between 0.5% and 1.5% of GDP annually for countries such as Costa Rica, Malawi, Mali and Mexico. In many areas bad irrigation practices have added to the problems, producing excessive salt residues in an estimated 24% of all arable land, and thus further reducing agricultural productivity.[19] For Africa as a whole, food production per capita has fallen steadily over the past decade.

---

[17] For more on this see e.g. Paul Harrison, *The Third Revolution*, Penguin, London 1992.
[18] *World Resources 1992-93*.
[19] G. Conway and E. Barbier, *After the Green Revolution*, Earthscan, London 1990.

domestic developments

Finally, population growth and the resulting pressure on the land have contributed significantly to the very fast growth of cities in developing countries. The proportion of the world's population living in cities has grown from 40% to 55% in the past 20 years, with a growth of over 1 billion in the numbers living in third world cities. A number of these cities (by the year 2000 Mexico City will have nearly 26 million people; Sao Paulo 24 million people; Calcutta 16.5 million people[20]) now dwarf most major cities of the developed world and have populations that are growing much faster than supporting infrastructure can be put into place. As a consequence parts of these cities tend to combine very heavy levels of overcrowding, poor sanitation, poverty, disease and other forms of environmental contamination.

The second group of developing country environmental problems, those brought by industrialization, are, unsurprisingly, similar to those that emerged in the West at a comparable stage of economic development.[21] Industry, mining and the growing use of agricultural chemicals are beginning to contaminate rivers and groundwater. Dangerously high accumulations of lead and other toxic metals have for example recently been found in rivers in Indonesia, Malaysia, Brazil, Korea and Turkey. By the late 1970s, 42 of Mexico's 65 major rivers were heavily polluted and another 16 moderately polluted. In Latin America seepage of toxic substances into groundwater reserves, mostly from waste dumps, is now estimated to be doubling every 15 years. Urban air quality in industrializing countries is now significantly worse than in the developed world, and the gap has widened over the past decade. In cities such as Bangkok, Beijing, Calcutta, New Delhi and Tehran concentrations of suspended particulate matter (SPM) exceed World Health Organization guidelines on more than 200 days a year and it has been estimated that excess SPM levels are responsible for 300,000-700,000 premature deaths a year in developing countries. While sulphur dioxide emissions in rich countries have fallen significantly in the past 20 years, they have risen sharply in industrializing countries (in Turkey they rose by a third in the late 1980s) so that more than

---

[20] Estimates are taken from N. Sadik, *Safeguarding the Future*, UNFPA, New York 1991.
[21] Statistics in what follows are drawn from the World Bank and D. Morell and J. Poznanski, 'Rhetoric and Reality: Environmental Politics and Environmental Administration in Developing Countries', in M.J. Leonard, ed., *Divesting Natures Capital*, Holmes and Meier, New York 1985.

1 billion people are exposed to levels well above WHO recommended limits, notably in Seoul, Bombay, Rio de Janiero, Calcutta, Hong Kong and Tehran. In Mexico City 29% of all children have unhealthy levels of lead in their blood, and in Bangkok blood lead levels are such that children are estimated to lose four or more IQ points before the age of seven. Mexico City schoolchildren in art class now reportedly paint the sky grey.[22]

Population

The first set of issues listed above, the pressures on land and water resources stemming from population growth and poverty, is of course not new. There is good evidence that such pressures – leading to deforestation and gradual destruction of the fertility of agricultural land – were a major factor contributing to the fall of the ancient civilizations of the Maya, the Sumerians and the Indus valley.[23] Plato described the effects of deforestation and soil erosion in classical Greece, and the Emperor Hadrian issued (unsuccessful) edicts to limit its impacts in Roman north Africa

Coming to more recent times, it has been plausibly argued that the degradation of the agricultural base and consequent impoverishment of the population have played a major role in provoking current political turmoil in such countries as Ethiopia, the Sudan, the Philippines and Central America;[24] and that similar factors are likely to play a growing role in Pakistan, parts of India and Bangladesh.[25] These are plainly crucial issues for the countries concerned, and also have wider implications for international stability – not least through the large numbers of 'environmental refugees' to which they have given rise. But their emergence, and the popular and governmental responses to them, have not in general been seen as a matter of environmental

---

[22] DuMars and Beltran del Rio, 'Air and Water Quality Laws of Mexico'.
[23] C. Ponting, *A Green History of the World*, Penguin, London, 1991.
[24] N. Myers, 'Population Growth, Environmental Decline and Security Issues in SubSaharan Africa', in A. Hjort of Ornos and M.A. Mohamond Salih (eds), *Ecology and Politics: Stress and Security in Africa*, Scandinavian Institute of International Studies, 1989; also T. Homer Dixon, Boutwell and Rathjen, 'Environmental Change and Violent Conflict', *Scientific American*, February 1993.
[25] N. Myers, 'Environmental Security: The Case of South Asia', *International Environmental Affairs*, Vol. 2, Spring 1989.

domestic developments 61

politics so much as of basic developmental strategy. They are not the sort of issues that have given rise to domestic environmental (as opposed to, for example, land reform or purely revolutionary) movements. Their impact on governmental environmental thinking in developing countries (as was made abundantly clear at Stockholm) has in general been to strengthen determination to industrialize at almost whatever price in pollution. It seems likely that they have had a similar dampening effect on the emergence of popular environmental movements in developing countries. It is difficult to argue strongly about air pollution when one's rural relatives live on the edge of starvation.

Developing countries have also begun gradually to tackle the problems posed by population growth[26] more directly. In fact the first developing country population policy, that of India in 1951, preceded by a considerable margin the outbreak of Western environmentalist concern on the subject. Wider acceptance of the idea of population control was initially very slow and uncertain. In China in the mid-1960s Mao argued that large populations were good for socialism. The 1968 papal encyclical *Humanae Vitae* bolstered opposition to population control in Catholic countries, so that as late as 1974 Argentina was arguing that many countries were suffering from a problem of under- rather than overpopulation. Nevertheless, a growing number of developing countries were establishing population programmes, which by 1970 existed in Egypt, India,

> **The first developing country population policy, that of India in 1951, preceded by a considerable margin the outbreak of Western environmentalist concern on the subject ... Developing countries have increasingly concluded that population planning can assist economic development .. the international rhetoric has not been helpful.**

---

[26] Johnson, *World Population and the United Nations*. In terms of numbers it is in developing countries that the population problem is now concentrated. Population growth diminishes with growing prosperity (subject to wide local variations stemming in particular from different cultural attitudes to childbearing and the division of labour among family members; see e.g. Tolba et al., *The World Environment 1972–1982*). In most developed countries reproduction rates are already close to, or below, replacement levels. Hence the theory of 'demographic takeoff' – once a country gets rich enough population growth falls.

Indonesia, Iran and Thailand as well as in two Catholic countries, the Philippines and Chile. China launched its vigorous and effective (but also highly controversial) programme in 1972 after the turbulence of the Cultural Revolution had subsided. Mexico incorporated a reference to family planning in its constitution in 1974. More and more countries have joined the movement, so that by 1994 some 70 developing countries, including well over three-quarters of the developing world's inhabitants, have formal population policies.

Not all of these policies have proved effective. They have made very little impact on fertility rates in Nigeria or Ethiopia for example. Nevertheless their global effect has been described by the UN, in a notable departure from its usual understatement, as a 'spectacular decline in fertility'. It has helped global fertility rates (the number of children born in the lifetime of each woman) to fall by about a third since the 1950s. Of course other factors, such as growing prosperity in various regions of the world and rising educational levels, have also helped (raising educational levels for women is the single most effective way of bringing fertility down); but even taking these into account it has been estimated that world population is now about 400 million (8%) less than it would have been without active population control policies and that on reasonable assumptions it will be about 4 billion smaller through the next century.[27] The middle UN estimate is that world population will stabilize at roughly double its present level by the end of the next century. This will obviously still entail a sharp extra strain on the world's resource base and natural systems. There are moreover particular regions of the world, notably sub-Saharan Africa, where the growth rate of population has continued to rise (from 2.5% in the early 1960s to 3% in the early 1990s) and where a catastrophic mismatch between numbers of people and the resources available to support them remains a very real prospect. Nevertheless, at the global level, earlier deeply alarming growth projections have been significantly cut back and the Malthusian/Ehrlichian nightmare of population growing exponentially until cut off by famine is no longer the mainstream prediction.

How has this progress been achieved? The growth of Western concern about the issue, with the subsequent mobilization of significant aid and technical assistance, both through the UN and bilaterally, has certainly played some

---

[27] *World Resources 1992-93*

part. Current (1993) aid for population projects is approaching $1 billion per annum.[28] A significant proportion of that sum is transferred through the UN and its population arm UNFPA, and this constitutes perhaps 40% of the total cost of population programmes in developing countries (excluding China, whose programme is almost entirely domestically financed). Moreover, the growing accumulation of objective data on the economically destructive effects of overpopulation, transmitted through the world's major bilateral and multilateral aid agencies, has undoubtedly had an effect. It is also likely that there has been a 'demonstration effect', as countries have learnt from successful programmes in other countries in their region (the most active areas for new programmes have moved in waves first from Asia to Latin America and now to Africa).[29] At the individual level most population policies (with conspicuous exceptions in China and Ceauçescu's Romania) have worked through giving people the choice (e.g. by making contraceptives available) of having fewer children, often backed up by education and encouragement (and, in the interesting case of Singapore, financial incentives). Once the socio-economic factors become right, people increasingly opt for the ecologically beneficial alternative because they judge it also to be in their own best interest.

Multilateral discussion of population issues, on the other hand, has been much less productive. We have already noted the division that prevailed at Stockholm, with Northern politicians under pressure to do something about the burgeoning millions of the poor and Southern politicians determined that discussion of Southern population growth should not be used to obscure the environmental damage done by high levels of consumption in the North. These divisions turned the first UN population conference (Bucharest, 1974) into a North-South confrontation in which the developing countries pressed for assurances of help in improving their economic position (on the entirely valid grounds that by becoming more prosperous their rate of population growth was likely to fall) and, having failed to get this, then removed all population targets from the plan of action produced by the conference.[30] The follow-up

---

[28] B. Crane, 'International Population Institutions: Adaptations to a Changing World Order', in P. Haas, R. Keohane and M. Levy, eds., *Institutions for the Earth*, MIT Press, Cambridge MA 1993.

[29] Johnson, *World Population*.

[30] G. Porter and J. Brown, *Global Environment Politics*, Westview, Boulder Co 1991; Johnson *World Population*.

conference in Mexico City in 1984 was even worse. Population policy had by then become entangled with Western (and particularly US) controversy about abortion, so that the US delegation at the conference declared population growth to be a 'neutral phenomenon' and followed this up by terminating funding to the UN Fund for Population Activities (UNFPA) and the International Planned Parenthood Federation (IPPF), the two major international population agencies.

The conclusion has to be that despite the political theatre at Stockholm, Bucharest and elsewhere, individual developing countries have increasingly concluded that population planning can assist their aims for economic development. These have often not been easy decisions to make. The Catholic countries of Latin America have reached them noticeably more slowly than the much more densely populated nations of South Asia. Many Muslim and African countries are still having trouble overcoming cultural, nationalist and other pressures in favour of population growth. The international rhetoric has, as we have seen, not been helpful. Population planning is thus a striking example of a policy area of high significance for the future of the global environment where individual national self-interest has firmly and successfully occupied the driving seat, taking quiet financial and technical help from the international community clustered in the back seat, but driving on undistracted by regular North-South rhetorical confrontations on the issue.

Industrial pollution

The bulk of that political activity in developing countries in the past two decades explicitly described by its practitioners as 'environmental' has focused on the impacts of industrialization. At the popular level there has been no sign of the epidemic of concern about pollution that hit the West in the late 1960s, but there has been a widespread pattern of protests against particular problems and projects. Examples abound. In 1980 citizen protests blocked the construction of a soda ash plant at Laem Chabang in Thailand, partly because it would damage local tourism.[31] In 1980, too, popular action blocked the construction of a paper mill in the Indian state of Madhya Pradesh on the

---

[31] Morell and Poznanski 'Rhetoric and Reality'.

domestic developments

grounds that the associated deforestation would destroy the lifestyle of the local indigenous people.[32] In the Indian state of Andhra Pradesh local villagers in the early 1980s compelled the closure of a local paper mill which was polluting the water and thus affecting local irrigation and fisheries.[33] India has also seen energetic local movements against particular industrial projects in Kerala, Goa and Uttar Pradesh.[34] Spain (then a developing economy) saw effective popular opposition in the early 1970s to the establishment of an aluminium plant in the north of the country. In Nigeria farmers have opposed local industrial development. Inhabitants of Mexico, Ecuador and other states have challenged local cement factories because of the effects of their pollution on local life.[35] Sao Paulo, Hong Kong and Mexico City have all seen widespread protests about air pollution and congestion.[36] Even China, not a country which in recent times has been hospitable to spontaneous public action, has seen a striking amount of environmental protest. Thus in 1979 alone Shanghai reported 339 incidents of confrontation between factories and the public over pollution, resulting in 49 full or partial shutdowns; and July 1982 saw, in effect, a three-day riot in Wuhan, when the local residents broke into a coal loading facility and damaged it extensively in protest at the dust clouds it was emitting.[37]

This pattern of local and issue-specific action has been broadly repeated in the emergence of environmental movements across the developing world.[38] In many developing countries the scope for the growth of such movements is in any case limited by such factors as authoritarian government and a lack of any tradition of grass-roots political activism. Nevertheless in a number of developing countries popular environmental movements have emerged. In India and Kenya, for example, hundreds of such movements have sprung up

---

[32] H. Leonard and D. Morell, 'The Emergence of Environmental Concern in the Developing Countries: A Political Perspective', *Stanford Journal of International Law*, 17, 1981.
[33] H. Leonard, 'Politics and Pollution from Urban and Industrial Development', in Leonard, ed., *Divesting Nature's Capital*.
[34] Leonard and Morell, 'Emergence of Environmental Concern'.
[35] Ibido.
[36] Ibido.
[37] Ross, *Environmental Policy in China*, Indiana University Press, Indiana 1988.
[38] For a survey see Nanita Yap, 'NGOs and Sustainable Development', *International Journal*, Winter 1989-90.

protesting about issues from tree felling to limestone quarrying.[39] The focus of these movements has tended to remain heavily local and focused on particular issues.[40] The emergence of nationwide environmental movements and green parties with a mass following and a coherent agenda on the Western pattern has been a much slower and sparser phenomenon.[41]

There have also been instances of environmental issues coming to play a role in the broader political processes of developing countries, though it is striking that this has tended to be confined to countries well on the way to industrialization. In the 1981 election in Greece, Andreas Papandreou made heavy use of Athenian concern about urban air pollution in his election campaign.[42] In Mexico, Miguel de la Madrid made environmental contamination a primary theme of his campaign for the presidency in 1982.[43] Kim Young Sam gave prominent emphasis to the need to combat pollution in the course of his ascent to the presidency of South Korea in 1992. In some countries the environment has become a 'carrier' in local politics for other, less publicly avowable, causes. One of the reasons why environmental groups sprang up in Eastern and Central Europe at the twilight of communism was that they provided a focus for those who were pressing for greater democratization. In Bulgaria, for example, Ecoglasnost became the core of the opposition movement which brought the communist regime down. A similar phenomenon is apparent in the Basque region of Spain, where environmental protest has become an important way to express Basque support for regional autonomy.[44]

But this broad developing country pattern of issue-specific protests and movements, coupled, in certain more industrialized developing countries, with odd incursions into the wider political arena, displays nothing like the force

---

[39] McCormick, *The Global Environmental Movement*.

[40] C. Thomas, *The Environment in International Relations*, RIIA, London 1992. Chapter 3 and the references cited there are good on the distinction between Northern and Southern environmental movements.

[41] It is worth noting, though, that national federations of NGOs have now emerged in Indonesia, the Philippines, Kenya and Mexico (Porter and Brown, *Global Environment Politics*) and that small green parties have come into existence, with some popular support in inter alia Egypt, Brazil and Mexico (W. Rudig, 'Green Parties around the World', *Environment*, Volume 33 No.8, October 1991).

[42] Leonard, 'Politics and Pollution'.

[43] duMars and Beltran del Rio, 'Air and Water Quality Laws of Mexico'.

[44] Leonard and Morell, 'Emergence of Environmental Concern'.

domestic developments

with which the environment burst on the Western political scene a few years earlier. Particular issues apart, developing country governments have not been under the same popular pressure as their Western counterparts to tackle environmental degradation. Moreover, in their statements at Stockholm they went to some trouble to emphasize that this was not an issue of primary importance for them.

It is striking therefore how much action, at least at what we have defined as response level two, they actually did take in the years following Stockholm. The number of governmental agencies with explicit environmental management responsibility in developing countries rose from 11 in 1972 to 102 in 1980,[45] while the number of ministerial-level departments of the environment worldwide rose from fewer than 10 in 1970 to over 100 in 1994, the majority of them in developing countries.[46] These new ministries, wherever they have been established, have tended to bring with them a trail of environmental legislation. By the end of the 1980s virtually all the 25 countries in the Latin American and Caribbean region had legislation covering the protection of water, forests, wildlife, soils, coasts, natural resources and sanitation, and six of them had progressed to explicitly environmental framework legislation.[47] The 1970s saw Malaysia introduce water pollution laws, mostly to tackle the pollution from the country's palm industry, which have been described as 'some of the most impressive environmental statutes in the developing world'.[48] In the same period Thailand introduced strict statutes to limit the headlong deforestation of the country,[49] and the Mexican government launched a major fiscal campaign to direct polluting industry away from Mexico City. Even Brazil, despite angry statements at the time of Stockholm about American 'cultural imperialism' with regard to the environment and dismissive references to American 'geolatry', had by 1976 introduced a range of antipollution policies and was beginning to reflect seriously on the problem of Amazon deforestation.[50]

---

[45] Ibid.
[46] Ausubel and Victor, 'The Environment since 1970'.
[47] Tolba et al, *The World Environment 1972–1992*.
[48] Leonard and Morell, 'Emergence of Environmental Concern'.
[49] Ibid.
[50] Morell and Poznanski, 'Rhetoric and Reality'.

The cases of India and China are especially revealing. Both countries had taken a firm third world line at Stockholm. Nevertheless, in both New Delhi and Beijing, Stockholm seems to have prompted a marked increase in governmental attention to environmental problems.

In the case of India, Indira Gandhi seems to have come away from the Stockholm conference determined to push forward India's environmental policies.[51] With her vigorous support, and despite negligible public interest, India adopted in 1974 a Water Act which has been described as the keystone of its environmental legislation. Mrs Gandhi herself fell from power in 1977, partly as a result of the excessive vigour with which she pursued her highly unpopular population control policies, but on her return to office in 1980 she carried the process forward with clean air and forest protection legislation and the establishment of a cabinet-level ministry of the environment. The process has rolled on since. By the mid-1980s India had nearly 50 pieces of environmental legislation, which were pulled together in a law passed in 1986 following the Bhopal accident. In the decade since the establishment of the environment department the number of its personnel has quadrupled and environmental spending by central government has risen from $40 million in the 1980-85 five-year plan to $357 million in the 1985-90 five-year plan, although it still constitutes only 0.2% of the total national budget.

China has followed a similar trajectory.[52] Following Stockholm the bureaucracy initiated a number of local schemes for environmental improvement and in 1973 produced the first ever State Council Directive on the subject. There was then an interregnum in which the future course of economic (including environmental) policy depended on the outcome of the power struggle between the radical 'Gang of Four' and the pragmatists led by Deng Xiaoping. With the triumph of the latter, environmental policy began to make significant advances. The first full statute on environmental protection was adopted in 1979 and provided for environmental impact assessments, sanctions, enforcement procedures and a system of effluent charges with fines for excessive discharges. Chinese environmental spending subsequently rose from 1.8

---

[51] see Khator, *Environmental Development and Politics in India*, for fuller details of the Indian case.
[52] see Ross, *Environmental Policy in China*, for more details.

domestic developments

billion yuan (0.25% of the Chinese equivalent of GDP) in 1980 to 5.4 billion yuan (0.5% of 'GDP') in 1984.

These histories point to a striking difference between environmental policy as it has emerged in developed countries and as it is emerging in developing countries. In the former, environmentalism was very much a 'bottom up' phenomenon.[53] Widespread popular concern rapidly communicated itself to democratically elected governments, and protection of the environment quite quickly became a consensual objective for public policy. In developing countries, by contrast, popular interest has been sporadic in its timing and highly local in its targets. Following Stockholm, however, key third world governments have begun to develop environmental policy from the top down. Despite widespread popular indifference, the government pursues environmental objectives because it believes them to be in the national interest.

The evidence strongly suggests that despite the conflictual rhetoric at Stockholm the conference marked a key stage in the communication to developing country governments of Western views on the importance of the environmental issue. Not only did the conference itself raise third world awareness of the problems, it also began to make the environment an active consideration in a whole range of other agencies which act as points of contact and influence between the developed and developing worlds. These include national aid agencies, multilateral funding bodies (the views of the World Bank have already been noted), scientific circles and other non-governmental organizations. There is also evidence that third world urban elites, who are often receptive to Western ideas and are exposed to burgeoning urban pollution, have played a key role in pushing developing country governments into tackling pollution problems in such countries as Mexico, India, Nigeria and Argentina.[54] The combined effect of these various channels of Western influence was summed up by a senior Spanish official in 1980 when he said: 'America is the leading country in the world. If Americans like to sing and dance everybody in Spain and in Africa and in Hong Kong sings and dances. The same is true with protests about nuclear power or pollution.'[55]

---

[53] Or, given the middle class origin we have already noted of Western environmental activism, at least 'middle-up'.
[54] Leonard, 'Politics and Pollution'.
[55] Quoted in Leonard and Morell, 'Emergence of Environmental Concern'.

This does not mean that developing countries are yet putting anything like the effort into environmental protection that developed countries are. They have remained true to their position that development comes first, at least as far as spending goes. Thus, while OECD country spending on the environment hovers around 1% of GNP or more, a World Bank survey of 40 developing countries in 1980 indicated that not one of them spent more than 0.3% of GNP on the environment.[56] Such patchy evidence as we have also strongly suggests that, while most developing countries now have some framework of environmental regulation, enforcement of those regulations, is often a problem. A 1976 UNCTAD survey of the rigour of environmental policy in 40 countries underlined its relative slackness in most developing countries at that time, with only Chile, Colombia, Israel and Singapore being classified as having relatively strictly implemented policies, and there are clear signs that the problem continues.[57] In India, for example, the inbuilt flexibility in the environmental laws combined with the inertia and, on occasions, corruption of the bureaucratic system have hindered the enforceability of tough standards so that pollution continues to grow and deforestation to increase.[58] In Mexico many pollution laws are ineffective because the fines are negligible and enforcement sporadic.[59] Malaysia has failed fully to enforce its impressive array of water pollution laws.[60] And in China, apart from the normal bureaucratic and corruption problems, industries confronted with a choice between achieving their planned production targets and meeting environmental regulations almost inevitably opt for the former.[61]

The overall picture that emerges of environmental policy in developing countries is of a top-down process which has begun to establish the necessary agencies and laws but in which popular pressures and enforcement mechanisms are not yet adequate to begin to turn around the rising tide of pollution. In the period after Stockholm many more countries found their way to a level two response, and successfully passed laws, but not to a level three response

---

[56] Morell and Poznanski, 'Rhetoric and Reality'
[57] World Bank, *World Development Report 1992*, Washington DC.
[58] Khator, *Environmental Development and Politics in India.*
[59] Morell and Poznanski, 'Rhetoric and Reality'.
[60] Ibid.
[61] Ross, Environmental Policy in China.

domestic developments

where those laws are effectively enforced and make a real difference on the ground.

The evidence also suggests, however, that as countries become increasingly industrialized the political and economic prerequisites for an effective attack on pollution, that is, for the shift to a level three response, begin to become available. The tentative emergence of the environment on the national political scenes of Mexico, Greece and South Korea has already been mentioned. Similarly, opinion polls in Sao Paulo and Taiwan have shown a rising level of generalized public concern about pollution.[62] There is evidence too that governments of the more industrialized developing countries are indeed beginning to make inroads on some of the problems.[63] Thus while average concentrations of suspended particulate matter and sulphur dioxide in cities in low-income countries (as classified by the World Bank) continue to rise, they have begun to fall in cities in middle-income countries. While river quality continues to deteriorate in low-income countries it has now stabilized in middle-income countries. UNEP figures suggest that sulphur dioxide concentrations have peaked and are now declining in some cities, including Caracas and Beijing. In the case of China (which despite its spuriously low estimated GNP is undoubtedly now a very rapidly industrializing middle-income country) a wide range of environmental indicators ranging from wastewater treatment rate to recycling rates are now improving and, with economic liberalization, the extensive and pioneering introduction of effluent charges shows signs of making some impact on the enforcement problem.[64]

4.4 The Soviet Union and Eastern Europe

We must complete our survey of environmental developments after Stockholm with a look at the formerly communist countries of Eastern Europe.[65]

---

[62] Morell and Poznanski, 'Rhetoric and Reality'.
[63] The figures that follow are drawn from World Bank, *World Development Report 1992*.
[64] Ross, *Environmental Policy in China*.
[65] Sources for this section include: World Resources 1992-1993, World Bank, *World Development Report 1992*, W. Beckerman, *Economic Development and the Environment*; T. Gutner, 'The Post Socialist Transition in Hungary and Czechoslovakia', Harvard mimeo; M. Feschbach and A. Friendly Jr., *Ecocide in the USSR*, Basic Books, New York 1992.

These countries were of course peculiar in the 1970s and 1980s in combining fully industrialized and state-run economies with authoritarian political systems. The environmental cost has been high. East European emissions of sulphur dioxide per dollar of GNP are typically five to ten times the figures for OECD countries. Sulphur dioxide concentrations in certain regions of Czechoslovakia have almost doubled since 1970. On two occasions in the mid-1980s in Prague sulphur dioxide levels rose to some twenty times the admissible 24 hour limit, in a manner comparable to the notorious London smog of December 1952. Budapest, too, regularly suffers from excessive levels of photochemical smog. Rates of forest defoliation are two to three times West European levels. The proportion of drinkable water in Poland's rivers fell from 33% in 1967 to 4% in 1976, and in Czechoslovakia less than 35% of river water is fit even for industrial use. In the former Soviet Union the unique ecology of Lake Baikal (the most voluminous and deepest fresh water lake in the world) is seriously threatened, while the Aral Sea has shrunk by 40% as a result of irrigation projects and is suffering from very high levels of contamination. The population of Central Europe and the former Soviet Union suffer from strikingly high levels of respiratory disease and childhood lead poisoning. Partly for these reasons life expectancy in the region is about five years shorter than in Western Europe and infant mortality, while much less than in developing countries, is high in comparison to the rest of the industrialized world. The estimated costs of clean-up are enormous, amounting for example in the case of Poland to $250 billion over the next 25-30 years (i.e. over 10% of current GNP per year).

There is no dispute about the main causes of this lamentable situation. The managers of state-owned factories were pressed to meet production targets rather than worrying about waste or emissions. The communist economic system dramatically underpriced certain key inputs, notably oil and coal, which were therefore extravagantly used (East European factories use about two to three times as much energy to produce a ton of steel as do those in Western Europe). Prices did not reflect the costs of cutting pollution so, again, industry had no incentive to do so, nor had it any incentive to introduce improved technology so as to reduce waste and raise efficiency.

These economic mechanisms operated despite the existence of legal frameworks intended to limit environmental degradation. As elsewhere in the industrialized world, the 1960s and early 1970s saw a growing awareness in Central Europe of the environmental consequences of industrialization. Despite the non-participation of the countries of the region in the Stockholm Conference, and their claim that environmental destruction was a consequence of Western capitalism, there was in fact quite extensive discussion of the issue in the region at that time, even accompanied by the emergence of weak environmental movements in Hungary and Czechoslovakia and a few instances of environmental protest. Most of the countries in the region introduced environmental standards (such as the Soviet Air Protection Law of 1981) and institutional systems intended to back them up, including in many cases antipollution fines.

But those systems plainly made little impact on rising levels of pollution. They tended to be weakly enforced since observance of the standards established in them depended upon on one (usually new and uninfluential) state agency effectively regulating other (in general much better established and powerful) ones. Such fees and fines as were charged were simply budgeted for in the sectoral plan so that factories should not be inhibited from maintaining production. And the popular pressures which in the West would have hindered such bureaucratic complaisance, and indeed which played a crucial role in compelling Western governments to take effective action to tackle pollution, had nothing like the same leverage under communism. In Russia for example, citizens and prominent intellectuals could write carefully worded letters to the press and join government sponsored conservation movements, but little else.[66] Direct criticism of the Party, and the broad thrust of government policy was not allowed. There was no free press or political opposition to help inform the public and to campaign on particular issues. Unsurprisingly, even on those issues (such as the pollution of Lake Baikal) where quite widespread concern was apparent it could not at that time make itself felt in a way that could compel the authorities significantly to change policy.

There is even evidence that in some cases East European governments consciously manipulated their domestic environments for economic or international ends. Thus, in order to meet requirements taken on under the Long

---

[66] P. Pryde, *Environmental Management in the Soviet Union*, Cambridge University Press, Cambridge, 1991.

Range Transboundary Air Pollution Convention (which was a major plank of East-West detente in the late 1970s and early 1980s) the Soviet Union actually moved polluting industry east, so increasing sulphur dioxide emissions by 12% in its Asian region. This enabled it to cut emissions in its European region, and thus the pollution blown westwards across its western border which it was required to cut under the convention. There is good reason to believe, too, that as part of its industrialization efforts of the late 1970s Romania specifically pursued the installation of a variety of chemical production processes which were banned or tightly regulated in the West for environmental reasons.[67]

The collapse of communism in the region was accompanied by a sharp rise in the public expression of environmental concern. This happened first in Poland where the official recognition of Solidarity in 1980 was followed by the establishment in the same year of the Polish Ecological Club the first fully independent environmental group in the region which formed part of a 'round table' intended to bring an end to communist rule. Later in the decade environmental movements played a major part in the wave of revolutions which brought the communist governments down. The role of Ecoglasnost in Bulgaria has already been mentioned. Another example is offered by Hungary, where the proposed Gabcikovo-Nagyamaros Dam on the Danube triggered the creation of the country's first environmental movement; public criticism of the dam project rapidly became synonymous with criticism of the regime, as it implied questions about the whole manner of governance in Hungary over the preceding 40 years. After the fall, public opinion polls revealed widespread concern about environmental degradation. In Czechoslovakia 83% of those questioned in one poll cited environmental questions as the most important problem facing the government.

However, this level of concern did not last. In the cold post-communist dawn populations and governments rapidly transferred their attention to the severe economic problems they faced and environmental issues moved rapidly down the public agenda. In both Hungary and Czechoslovakia, for example, weak environmental ministries found themselves facing powerful

---

[67] Leonard, 'Politics and Pollution'.

domestic developments

finance ministries,[68] and governments initially devoted their attention almost exclusively to economic liberalization, privatization and industrial restructuring. NGO and green movement numbers collapsed as many of their members, having used the movements as a foil to bring down the old regimes, then moved on to jobs in government and politics. This of course underlines the experience of OECD and developing countries that populations require a certain level of economic security and prosperity before they are likely to devote a significant amount of attention to environmental quality (other than as a surrogate for other issues, such as political freedom).

4.5 Agriculture

I will close this chapter with a look at two areas where the evolution of domestic environmental policy over the past twenty years has notable lessons to teach. The first is a sector whose environmental impacts in the developed and developing worlds have been closely (and pathologically) interdependent. This is agricultural policy. Although agriculture constitutes a relatively small and diminishing proportion of world GNP (about 10% in 1965, less than 7% now) it directly shapes about one-third of the world's land area (excluding Antarctica) and in the poorer countries continues to constitute a very significant proportion of economic activity (17% of GNP and about 60% of employment).[69] We have already seen that agricultural concerns played a large part in the early growth of Western environmentalism for two (mutually contradictory) reasons. The first was the widespread fear, as expressed by the Club of Rome, that world food production would not be able to keep pace with population growth. The second was alarm (as expressed, for example, in *Silent Spring*) at the pollution and dislocation of nature caused by the agricultural techniques and substances increasingly deployed to raise production so as to meet the pressures of increased population.

---

[68] A situation which is of course also familiar in the West but is (to some extent) palliated by the evolution of strong public support for action to tackle environmental degradation. It is however arguable, and consistent with the institutional emphasis of the Brundtland report (qv) that the only real solution to the problem of environmentally indifferent finance ministries is the full incorporation of environmental costs in the national accounts.
[69] Ibid.

We have also seen that the first of these fears was misplaced. In the period 1970-90 agricultural output rose 3% per annum in developed countries and 2% per annum in developing countries. Overall the world produced 22% more food in the 1980s than in the 1970s which, when population growth is taken into account, amounts to a 4% increase per capita.[70] Of course this progress was not evenly distributed. In Africa, uniquely, food production per capita has actually fallen over the past decade; and there are both international and intranational inequalities of distribution so that in practice about 10% of the world's population, mostly in Asia and Africa, is classified by the World Food Commission as 'chronically hungry'.[71] But if consumption were evenly distributed, world food production is enough to supply adequate nutrition for all.[72] Nor, despite the resolutely pessimistic tone of some writing on the subject, is it likely that this situation will change in the near future.[73] The biotechnology revolution continues apace.[74] There is still enormous scope for modernization and intensification of agriculture in the developing world, for example through increasing use of fertilizer.[75] Improved management of formerly communist agricultural sectors can be expected to bring sharp production gains. A rise to Western productivity levels of cereal production in the former Soviet Union alone would raise total world output by 10%. Indeed, the supply/demand balance (in what is admittedly, as we shall see below, a very imperfect market) has brought about a drop of a third or more in the real prices of agricultural produce over the past 20 years.[76]

---

[70] Ibid.

[71] M. Tolba et al., *The World Environment 1972–1992*, Chapman and Hall 1992. Dreze and Sen, *Hunger and Public Action*, Oxford University Press, 1989, give a very good account of the social and economic inequalities which lead to such deprivation and the success of some countries, including relatively poor ones, in tackling it through public provision.

[72] Ibid.

[73] See e.g. Lester Brown et al., *State of the World 1993*, Earthscan, London 1993. Dreze and Sen (ibid), after looking at the various available studies, have dismissed such shrill announcements of impending disaster and doom.

[74] A good example is the current gradual entry into use of bovine growth hormone which can raise milk production by about a third (Economist, biotechnology survey, 30 April 1988).

[75] World Bank, *World Development Report 1992*.

[76] Ibid. This overall conclusion is also strongly borne out by the FAO's famous 1982 study of global carrying capacity, Graham Higgins et al., *Capacities of Lands in the Developing World*, FAO, Rowe 1982.

The price for adequate nutrition for all (or most) as world population continues to rise must evidently be continuing growth in the productivity of world agriculture through the spread of modern agricultural organization and methods. Thus it was *Silent Spring* that identified the environmental problem more accurately. The intensification and extensification of agricultural production, the development and dissemination of high-yield crop varieties, and increased application of fertilizers and pesticides which have brought about the rising production of the past 20 years have carried significant environmental costs. Although expansion of arable land has accounted for a relatively small proportion of the growth in production, it amounted to a 5% increase in cropped areas worldwide from the 1970s to the 1990s, and a 9% rise in the developing world. This has inevitably brought agriculture to more marginal lands where the risk of, for example, soil erosion is greater and has been achieved by the destruction of natural habitat. Sixty per cent of deforestation in developing countries is estimated to result from agricultural expansion.[77] Intensification of cultivation has in many areas led to soil loss and degradation. Almost 11% of the world's total vegetated surface – one-third of its total agriculturally employed surface – is estimated to have experienced some loss of fertility.[78] Similarly, the rapid spread of monoculture and high-yield crop varieties has taken a considerable toll of the diversity of strains upon which we have hitherto been able to draw in breeding new varieties to meet new pests and other problems. It is estimated that by the turn of the century 12 rice varieties will account for 75% of the crop in India, where 30,000 varieties were to be found in 1980.[79] The 100% rise in world pesticide and fertilizer use in the past 20 years (much of it in the developing countries) has led to rising nitrate concentrations in groundwater (which now exceed EC recommended levels on about 25% of the land area of the European Community) as well as the doubling in the period 1970–90 of the number of resistant pest species and a build-up of pesticides in the food chain in a variety of developing countries. Breast-milk samples in cotton-growing areas of Nicaragua and Guatemala have revealed some of the highest levels of DDT ever recorded in humans, and the illness and death rates in such regions from pesticide poisoning approach those from

---

[77] Tolba et al, *The World Environment*; World Bank, *World Development Report 1992*.
[78] World Bank, *World Development Report 1992*.
[79] Tolba et al, *The World Environment*.

major diseases.[80] Pesticide residues have also been found in the groundwater of 34 US states.[81]

While all of these problems are apparent in both developed and developing countries, it is possible to draw a very broad distinction between their incidence in the two regions. In the developed world the key problems, as with industry, are the waste streams and byproducts of a highly intensive and technological agriculture – notably pesticide residues and nitrate runoff. There are problems of erosion and habitat degradation, but they are of a lesser order of magnitude. In the developing world this situation is reversed. While there are regions suffering from excessive pesticide and fertilizer use,[82] the key agricultural problems in developing countries stem from poverty and the consequent need to 'mine' the natural resource base to achieve maximum output now.[83] Such mining manifests itself in the form of habitat destruction, overcropping and neglect of long-term investment in the interests of present production (compelling examples of which are the destruction of antierosion terracing in Java to increase the current cassava crop,[84] and the increased cutting of trees for fuel in sub-Saharan Africa when income falls).[85] This is an important difference to bear in mind. Very broadly speaking (and with innumerable local caveats and variations), the environmental problems flowing from agriculture in the North stem from too much intensification and artificial inputs, whereas in the South they stem from too little.

Governmental responses to these problems have been a function of the very peculiar politics and economics of world agriculture. This is an economic sector which, in the manner of the former Soviet economy, is in most countries completely dominated by government planning and price-fixing. But here too there is a sharp dichotomy between the general approach in devel-

---

[80] World Bank, *World Development Report 1992*.
[81] *World Resources 1992-93*, World Bank, *World Development Report 1992*; Tolba et al., *The World Environment*.
[82] W. Beckerman, *Economic Development and the Environment*.
[83] World Bank, *World Development Report 1992*. For example, the FAO (*The State of Food and Agriculture*, 1991) estimates that in sub-Saharan Africa soil nutrient depletion due to overuse amounts to about 20 kg of nitrogen, 10 kg of phosphorous and 20 kg of potassium per hectare per year on average.
[84] Conway and Barbier, *Beyond the Green Revolution*.
[85] Beckerman, *Economic Development and the Environment*.

domestic developments

oped and developing countries. In most developed countries, and in particular in the US, the EC and Japan, agriculture is massively subsidized to a total of about $250 billion per annum[86] – more per year than the total of all World Bank lending since 1980. These extraordinary subsidies, whose ostensible justification tends to be the need to avoid rural depopulation, are maintained through the very powerful political influence of agricultural interests in Western polities. Their influence is demonstrated by the facts that about two-thirds of the entire EC budget is spent on agriculture, and that transatlantic disagreements over agriculture (which constitutes less than 10% of world trade) have undoubtedly been the most serious threat to efforts in the early 1990s to reform the world trade regime. The main consequence of the system in developed countries has been to hold domestic agricultural prices and production at a far higher level than would otherwise be the case, often leading to overproduction and the dumping of the excess on world markets.

The growth of Western popular environmental concern from the 1960s on has, as in the industrial sector, made some impact on the more obvious environmental damage caused by Western agriculture. The power of the farm lobby in the US, Japan and the EC is such that the most obvious approach – to cut the artificially high prices paid for agricultural produce and hence the amount produced and the environmental damage that goes with that production – has not been effectively pursued. Less direct routes have had to be used. Pesticide controls have been tightened: in the US pesticide use fell by about 20% from the mid-1970s to the mid-1980s (with a consequent recovery of wild bird populations).[87] Controls on fertilizer and manure application are also slowly being tightened in the EC to tackle the problem of rising nitrate levels in groundwater. As farm productivity, and so overproduction of farm produce, have continued to rise, governments are increasingly paying farmers to reduce production, often with an ostensibly environmental purpose. Farmers are represented as 'environmental stewards' of the countryside and their subsidies justified accordingly. Thus the EC has a system of 'environmentally sensitive areas' where farmers are paid to use traditional

---

[86] O. Knudsen, J. Nash, et al., *Redefining the Role of Government in Agriculture for the 1990s*, World Bank, Washington DC, 1990.
[87] *World Resources 1992–93*.

(low-productivity) agricultural techniques, and the US has paid farmers to take some 14 million hectares of highly erodible land out of production.[88] Rather more consistently with the 'polluter pays principle' that those who damage the environment should pay for that damage, a number of countries (notably Australia and Sweden, whose farm lobbies are significantly weaker than those in the EU and the US) have experimented with raising the prices of, for example, pesticides, fertilizer and water, leading to encouraging increases in the efficiency and specificity with which these inputs are deployed (another example of environmental protection through economic incentive rather than regulation).

Unlike in the developed world, domestic agricultural prices in developing countries have tended to be held below world prices. This approach, which has been significantly reinforced by the dumping of Western surplus agricultural produce on world markets, has reflected the crucial political importance in most developing countries of keeping urban food prices down. These low prices tend to be fixed and enforced through state marketing organizations.[89] Inevitably the consequent impoverishment of the countryside has contributed to urban migration and the growth of the developing country urban environmental problems that we have already noted. It has also had another pernicious effect. By lowering land values, and farmers' incomes, it has increased the pressure on farmers to maximize current production with little thought for, or incentive to invest in, the longer-term fertility of the land (while conversely, the artificially high prices of produce in developed countries have raised land values and farm incomes, and have encouraged farmers to maintain the productivity of their land).[90] It is thus likely to have contributed to the very high levels of land degradation in, for example, Asia and Africa, where about 20% of all vegetated land has undergone some degradation.[91] Develop-

---

[88] Ibid.
[89] Knudsen and Nash, *Redefining the Role of Government*.
[90] This is a dramatic simplification of a much more confused picture on which evidence is scanty, but property regimes and tenure arrangements also play a key part in the farmers' attitude to his land. See Conway and Barbier, *Beyond the Green Revolution*; D. Pearce, ed., *BluePrint 2: Greening the World Economy*, Earthscan, London 1991; Knudsen and Nash, *Redefining the Role of Government*.
[91] *World Resources 1992–93*.

domestic developments

ing country governments, whose policy aim has tended to be 'food security' – again with urban concerns predominant – have often tried to compensate their farmers for underpricing their produce by underpricing their inputs as well. Agrochemicals and irrigation are often very highly subsidized in developing countries (typical fertiliser subsidies vary from 50% to 95% of the market price).[92] In some ways the environmental consequences of this have probably been beneficial. Subsidized fertilizer use goes some way to offset the 'soil nutrient deficit' resulting from overcropping in many developing countries and thus helps limit the pace of land degradation. But in others it has not. Underpricing of water has led to massive waste and overuse with, for example, 10 million hectares of land in India lost through waterlogging and another 25 million threatened by rising salt levels.[93] Perhaps the most compelling example of environmentally damaging subsidization of agriculture was the incentives offered by the Brazilian government up to 1990 to encourage farmers to move to Amazonia. This opened the way to a 'slash-and-burn' style of shifting cultivation which, in that particular context,[94] has been described as ' the most environmentally destructive in the world'.[95]

The state planning of the agricultural sector which prevails in most countries is, however, now under threat. Cuts in Western agricultural subsidies have become part of the world trade regime. Both the US Congress and the EC Council of Ministers have worked (so far fruitlessly, but with growing intensity) to cut the inexorably rising costs of keeping their farmers in the style to which they have become accustomed. The collapse of communism has encouraged a worldwide shift to greater reliance on the market (ironically most apparent in one country where communism has not collapsed, China, which by introducing market incentives increased food production per capita by one-third in the course of the 1980s).

It is therefore a question of more than theoretical interest whether liberalization of the world agricultural market would benefit the global environment

---

[92] Knudsen and Nash, *Redefining the Role of Government*.
[93] Ibid.
[94] This should not be confused with traditional shifting agriculture, where the cultivator allows the land long fallow periods for regeneration and thus does not constitute anything like the same environmental scourge; see Thomas, *The Environment in International Relations*.
[95] Conway and Barbier, *After the Green Revolution*.

or not.[96] The most recent analyses suggest that the fall in developed country production (ranging from 12% for grain up to 40% for sugar) which would result from the abandonment of subsidies would be more or less offset by the rise in production in developing countries which would result from governments raising to world levels the prices they pay to farmers for agricultural produce.[97] Overall, world prices would change little, so that even the very poor food-importing developing countries (notably in sub-Saharan Africa) would not be greatly disadvantaged. It has been estimated that the benefit to developing world farmers as a whole would be of the order of $70 billion per year.[98] This is a significant sum, amounting to 12% of total current developing world agricultural GNP – substantially more than the West's *total* current annual aid to the developing world.

There would be environmental costs. A proportion of the world's agricultural production would move from the West, where farmers have access to the most up-to-date environmental protection technologies and training, and are subject to a reasonably firm framework of environmental regulation, to the South, in much of which neither of these conditions applies. Inevitably there would be disruption and environmental damage as traditional Southern patterns of land tenure and subsistence farming adapted to the production of cash crops for export.[99] Higher producer prices in developing countries would undoubtedly increase the incentives for overcropping, forest clearance and the farming of

---

[96] See Goldin and O. Knudsen, eds., *Agricultural Trade Liberalisation: Implications for Developing Countries*, Paris, OECD and World Bank 1990. The best analysis of the environmental impacts is K. Anderson, 'Effects on the Environment and Welfare of Liberalising World Trade, the Cases of Coal and Food', in Anderson and Blackhurst eds., *The Greening of World Trade Issues*; Harvester Wheatsheaf, London 1992.

[97] Knudsen and Nash, *Redefining the Role of Government*; Anderson, *The Greening of World Trade Issues*.

[98] Knudsen and Nash, *Redefining the Role of Government*.

[99] Although there is no reason to believe that cash cropping is as environmentally or economically objectionable as it has sometimes been painted. See e.g. Conway and Barbier. *Beyond the Green Revolution*, and Dreze and Sen, *Hunger and Public Action*, who give a long list of developing countries which have advantageously encouraged a shift to cash crops. There are of course real costs – social, economic and environmental – associated with unequal land tenure patterns and other sources of rural deprivation and insecurity in developing countries. But these do not seem to be intrinsically related to a shift to cash cropping (Dreze and Sen, *Hunger and Public Action*).

erodible marginal land, thus probably accelerating the destruction of some of the world's most vulnerable and biologically diverse ecosystems.

It is nevertheless difficult to resist the conclusion that these effects would be more than offset by the environmental benefits.[100] We have already noted that the principal environmental problems confronting developing country agriculture stem from poverty, and the consequent need to mine the resource base. A net transfer to developing countries of some $70 billion per year – concentrated on third world farmers, who constitute the vast majority of the world's poor – would amount to a very substantial boost to their prosperity. It could be expected to yield significantly greater attention to protection of the natural resource base from which the prosperity flowed. The higher income would permit a shift in emphasis from immediate production, at no matter what ecological cost, to longer-term attention to, and investment in, the productivity of the land.[101] Moreover, it seems highly plausible that moving crops from where they are most highly subsidized to where they grow best must reduce the overall strain on the global ecosystem and would thus constitute 'an important step in the evolution of sustainable agriculture'.[102] It has in particular been argued that global use of agrochemicals would fall dramatically as highly chemical-intensive Northern producers were replaced by more pastoral enterprises in areas such as Africa.[103] Last but not least, the relative rise in rural prosperity would help slow the drift to the towns, with all of the environmental pressures that is creating.

Thus liberalization of world agricultural markets offers a global environmental gain as well as the significant increase in economic prosperity in both developed and developing countries which would result from removal of the huge deadweight cost of the current distortions of the global agricultural market.[104] No altruism, or even environmentalism, is needed; just a clear-headed pursuit by the world's major countries of their own best economic interests.

---

[100] This conclusion is certainly not uncontested in the literature. But the logic seems to me to be inescapable. Those who argue that the current system of agricultural protectionism benefits the environment in effect argue (against a lot of evidence) that third world poverty does so too.
[101] For other arguments pointing in the same direction see, J. MacNeill, P. Winsemius and T. Yakushiji, *Beyond Interdependence*, Oxford University Press, Oxford 1991.
[102] *World Resources 1992–1993*.
[103] Anderson, *The Greening of World Trade Issues*.
[104] Knudsen and Nash, *Redefining the Role of Government*.

4.6 Freedom and the environment

The second area for which the history of the past two decades has rather clear implications is the relationship of environmental protection and political freedom. We noted in Chapter 2 the 'dark green' insistence that radical changes to the whole Western way of life are needed to respond to the environmental challenge. This has given rise to the contention that the environment can only be protected by a sharp curtailment of individual and economic liberties. Thus Ophuls wrote: 'because the free play of market forces and individual initiative produces the tragedy of the commons the market orientation typical of most modern societies will have to be abandoned.'[105] Similarly, Heilbronner has claimed that 'the latitude for voluntary choice may vanish rapidly under an avalanche of regulations and regimentation in response to the slowly closing vice of environmental constraint.'[106] More recently, Caldwell has suggested the introduction of 'crimes against nature', a category chillingly reminiscent of Stalin's 'crimes against the state'.[107]

This line of argument does not have the support of all environmentalists. It has been more common to hear dark green argument for anarchistic and decentralized government than for active coercion. But those who have advocated this approach have signally failed to reconcile it with their demands for stringent economic regulation to guarantee environmental protection. Such regulation would in effect amount to state planning of resource use and market operation,[108] and point the way (however unintentionally on the part of its advocates) to political authoritarianism as well.[109] Indeed, the logical link between dark green environmentalist ideas and coercive political structures is hard to evade. If people are allowed free choice then it seems clear that most of them will continue to opt for the car, the television and the fridge, even if

---

[105] W. Ophuls, *Ecology and the Politics of Scarcity: Prologue to a Political Theory of the Steady State*, W.H. Freeman, New York 1977.
[106] R. Heilbronner, 'Second Thoughts on the Human Prospect', *Challenge*, May, June 1975.
[107] L. Caldwell, *Between Two Worlds*.
[108] As noted by Dobson, *Green Political Thought*.
[109] For as R. Dahl (*After the Revolution? Authority in a Good Society*, 2nd edn., Yale University Press, New Haven 1990) notes 'It is a historical fact that modern democratic institutions... have existed only in countries with predominantly privately owned, market oriented economies. It is also an historical fact that ... centrally directed economic orders ... have not enjoyed democratic governments but have in fact been ruled by authoritarian dictatorships'.

domestic developments

they are prepared to pay marginally more to limit the polluting effects of these choices. Thus the only way to limit such materialism (and the ecological catastrophe to which it will supposedly lead) is by some sort of central economic regulation pointing exactly towards the authoritarianism that Ophuls and others anticipate.[110]

While it is impossible to deny the theoretical possibility of the state imagined by Ophuls and others in which individual liberty would be sharply circumscribed for the greater ecological good, it is possible to test that picture against those mechanisms which in the real world have promoted environmental protection, as well as those that have damaged it. Such a test does not lend much support to the coercive view of environmental protection. From an examination of the cases set out above it is quite clear that the impulse for environmental protection has been very much a democratic phenomenon. It has required freedom of association, freedom of speech, freedom of publication and a system of government responsive to public pressure to operate at its most effective. Moreover, the price signals provided by free markets have produced a much less wasteful and environmentally destructive allocation of resources than has been achieved in circumstances (such as the world's agricultural markets) where they have not been available. These channels for public and consumer pressure have been necessary to counterbalance the producer pressure which otherwise tends to dominate governmental decision-making and which, as we have regularly seen, tends to be deployed in opposition to environmental regulation. Happily these freedoms have been available in that part of the world where environmental protection has gone furthest, the democracies of

> **The impulse for environmental protection has been very much a democratic phenonemon. It has required freedom of association, freedom of speech, freedom of publication and a system of government responsive to public pressure to operate at its most effective.**

---

[110] Authoritarianism was of course also the outcome when in Russia in 1917 a party came to power on a utopian political prospectus and expectations of revolutionary changes in human values.

the West. Where they have been absent on the other hand, as until recently in Eastern Europe, it has been quite impossible for any popular concern about environmental degradation to exert effective pressure upon the authorities – and those authorities have shown very little spontaneous concern for the environment in the absence of such pressure.[111] We have also seen that the same has been largely true in the one sector of the Western economy, agriculture, where producer interests have become almost impregnable to economic, let alone environmental, pressures for a rational allocation of resources.

A less visibly disastrous example of the alleged environmental benefits of compulsion has been China's controversial 'one child per family' population policy, which has received some environmentalist support and exoneration.[112] Before welcoming such policies as being good for the planet (however unpleasant for those directly affected)[113] it has to be asked whether the same results could not have been achieved by more civilized means and whether they are likely to prove sustainable in the long term. It is striking, for example, that Sri Lanka, with a GDP per capita very comparable to that of China, has achieved an even more substantial fall in the rate of population growth over the same period. India's short flirtation with coercive population control not only brought down the government in 1977 but led to a 'bounce back' in India's birth rate which has taken it years to reverse.[114] More generally we have seen that in a wide variety of developing countries fertility rates have been successfully brought down and kept down through entirely voluntary methods. It must therefore be very questionable whether the loss of personal freedom entailed by the Chinese approach is necessary for, or justified by, the ecological gains, and how sustainable those ecological gains are likely to prove.

---

[111] Another supporting instance (which also suggests that right-wing dictatorships can be as environmentally undesirable as left-wing ones) is provided by Cubatao in Brazil where, following the reintroduction of democracy in 1983, the newly elected governor launched a massive environmental clean-up, cutting typical pollutants by up to 90% and reintroducing fish into the river from which they had been absent for 30 years (Harrison, *The Third Revolution*).

[112] Johnson, *World Population*.

[113] According to the Chinese planning agency for example, there are now 36 million more men than women in China, a figure that will rise to 70 million by the end of the century. It has been widely suggested that this discrepancy results from the determination of many families that the one surviving child they are allowed should be a boy.

[114] Harrison, *The Third Revolution*.

domestic developments

Thus, in the absence of persuasive examples of cases where coercion has been of benefit to the environment, and given the plenitude of examples where authoritarianism and government suppression of market signals have demonstrably damaged it, it is difficult not to conclude on the evidence currently available that the best way we have so far been able to identify of protecting the *general* interest – of which the environment is surely a part – is through the maintenance of a very large area of individual political and economic freedom.

Chapter 5

From Stockholm to Rio: International Developments

> *There can be no greater error than to expect, or calculate upon, real favours from nation to nation.*
> GEORGE WASHINGTON

5.1 Overview

The boom in developed country environmental concern which prompted the Stockholm Conference also produced a boom in the negotiation of international environmental treaties. This pattern is set out in Table 5.1. The number of significant international environmental agreements achieved in the 1960s was almost double that for the 1950s (with the bulk of the activity concentrated in the second half of the decade), and the number more than doubled again in the 1970s. The 1980s saw a slight falling off from the 1970s peak – but the pace of international environmental law-making remained more than three times that of the 1950s. Second, and less predictably, the crude numbers suggest that the bulk of effort was devoted to marine pollution, which accounted for slightly less than half of all the treaties negotiated both over the whole period and in the two decades following Stockholm. A third striking feature is the growth of regional agreements from a very small minority of the whole in the 1950s and 1960s to two-thirds of all major environmental treaties negotiated in the 1980s. This explosion in regional environmental business has gone rather unnoticed by comparison with the attention that has been focused on global issues. Finally, the figures hint at what are arguably the two most important facets of recent environmental treaty-making – the growing number of international agreements on the atmosphere, and the growth

Table 5.1 Multilateral agreements with significant sections on the environment, 1950–1989

	1950s	1960s	1970s	1980s	Total
Marine	4	4	14	8	30
Atmosphere	0	0	1	4	5
Wildlife/habitat	1	1	3	2	7
Nuclear	0	4	1	2	7
Antarctic	1	0	1	2	4
General framework	0	0	3	2	5
Other (e.g. space/waste)	0	2	3	1	6
Global total	5	9	12	7	33
Regional total	1	2	14	14	31
Total	6	11	26	21	64

Note: The category 'multilateral agreements with significant sections on the environment' is plainly to some extent arbitrary. It excludes bilateral agreements, major multilateral agreements which have indirect environmental effects (such as various GATT instruments) and international agreements with 'ornamental' references to the environment (such as UNCLOS I). It includes significant regional environmental agreements (such as the 1971 Scandinavian agreement to cooperate to tackle oil pollution) as well as general treaties with significant environmental passages (such as UNCLOS III and the European Community Single European Act). A major distortion is that it excludes major European Community environmental directives, which in any other context would undoubtedly be classified as significant environmental treaties. They have been adopted in such numbers, however, that if included they would swamp the other data. In general I have tried to be exclusive rather than inclusive, placing some weight on the word 'significant'. This undoubtedly lowers the totals for later decades and raises them for earlier, as what was significant in the 1950s had become commonplace in the 1980s. Data are drawn from the UN treaty series and the annual reports of the US Council on Environmental Quality.

of significant environmental passages in wider international agreements such as the European Community Treaties and the North American Free Trade Area.

This chapter does not pretend to cover all of the international environmental business of the period. It focuses rather on the most significant items, starting with the proliferation of agreements on marine pollution (with a separate section on those on the Mediterranean). It then covers the most significant conservation agreement to date, the Convention on the International Trade in Endangered Species (CITES), and the first major agreements on international atmospheric problems, those on acid rain. There follows a section on the most important forum for regional environmental cooperation, the European Community (now Union), and a look at one area where international cooperation might have been expected to develop but did not: nuclear safety. Finally, there is a swift survey of other international activity in the period which had implications for later environmental developments.

5.2 Marine pollution

We have seen that by the time of Stockholm there was already something of an international industry producing agreements on pollution of the oceans, notably by oil. This activity had been given a significant boost by the *Torrey Canyon* disaster. Developed countries with significant maritime interests now had civil servants and scientists with substantial experience in crafting such agreements, and the IMO had a long track record of organizing such negotiations in a politically uncontentious and technically proficient way. The agenda of the IMO was moreover dominated by the developed countries, who predominated in maritime terms and whose environmental concerns therefore received early attention. Those developing countries with growing fleets (notably Brazil, India, Mexico, Argentina, Indonesia and Nigeria) tended to be cautious towards environmental proposals which might hinder that growth; but they could not block discussion entirely, and in any case had the option of non-participation in any instruments it produced.

This area, in contrast to many others, thus had a well-developed international infrastructure conducive to further work after Stockholm.[1] Accordingly, the years after the Stockholm Conference added to a web of marine pollution agreements, which are summarized in Table 5.2. The story is complicated but important. It was in many ways in the field of marine pollution that the international environmental negotiating community learnt its craft. The interplay between global and regional agreements, the steady expansion of the number and sources of pollutants addressed, and the emergence of

---

[1] Accounts of the various agreements and their backgrounds can be found in E. Gold, *Handbook on Marine Pollution*, Assuranceforingen Gard, Rotterdam 1985; Churchill and Lowe, *The Law of the sea*, 2nd ed, Manchester University Press, Manchester 1988. For the developed/developing country balance in the IMO see M'Gonigle and Zacher, *Pollution, Politics and International Law: Tankers at Sea*, University of California Press, Berkeley 1979.

Table 5.2: Key marine pollution agreements, 1954–1983

Date	Agreement	Area of application	Comment
1954	International Convention on Oil Pollution	Global	Objective to cut oil discharges from tankers
1969	Three agreements to permit intervention, allocate liability and award compensation for tanker accidents on the high seas	Global	Reaction to *Torrey Canyon* disaster
1969	Bonn Convention	North Sea	To limit oil pollution of North Sea (reaction to *Torrey Canyon*)
1972	Oslo Convention	NE Atlantic and North Sea	To control marine dumping in the north-east Atlantic and North Sea
1974	Helsinki Convention	Baltic Sea	To control all sources of Baltic pollution
1974	Paris Convention	North Sea	To control all land-based sources of pollution of the North Sea (became the main basis for the North Sea Conferences 1984, 1987 and 1990)
1975	London Dumping Convention	Global	To control maritime dumping globally
1975/6	Mediterranean Action Plan	Mediterranean	Basis for joint action to tackle Mediterranean pollution
1980	Mediterranean Plan, Land-based Protocol	Mediterranean	To limit land-based sources of pollution of the Mediterranean
1983	International Convention on the Prevention of Pollution from Ships (MARPOL)	Global	To control all sources of maritime pollution from ships; very slow to enter into force; subsumes the 1954 Convention on Oil Pollution

international developments

dedicated international environmental funding, regular rounds of treaty amendment and the use of political conferences to break bureaucratic logjams, all originated in this field.

The Stockholm conclusions referred to, and encouraged, two extensions of the then existing international marine pollution regime. The first of these was the extension to global application of the Oslo Convention, which controls the dumping of wastes in the North Sea. This was quickly done in the so-called London Dumping Convention of 1972 (although its supposed global scope is rather undermined by the fact that so far fewer than 30 developing countries have ratified it, a group which does not include such key states as India, Argentina or Egypt). The second was continuation of the line of work dating from the original 1954 International Convention on Oil Pollution. This treaty, as we have seen, had already been strengthened a number of times in its history. The aim now, however, prompted by Western public alarm about the state of the oceans, was to extend its coverage from just oil discharges by ships to discharges of *all* polluting substances. This was done in the 1973 International Convention on the Prevention of Pollution from Ships (MARPOL).

> It was in many ways in the field of marine pollution that the international environmental negotiating community learnt its craft.

MARPOL was an ambitious document. It does indeed extend the old treaties on oil pollution to a much wider range of substances, and requires the installation of expensive new antipollution technology in all new tankers. It has therefore been described as 'the most important and comprehensive treaty to fight marine pollution'.[2] But its subsequent fate firmly underlines how limited is the scope for rapid clean-up of the international commons through bold treaty-making. MARPOL was adopted largely because of support from states with coastlines long enough to be worried about marine pollution, and shipping industries small enough to be undeterred by the costs (such as Australia, Canada, Argentina and India). But precisely because MARPOL's provisions were so wide-ranging and costly, states with large tanker fleets, such

---

[2] B. Moslam, 'Global Marine Pollution Treaty has been Ratified', *World Environment Report*, 8, 30 November 1982.

as Greece, France and Japan, were very slow to ratify it – so much so that by 1978, five years after the convention was signed, it was clear that if it was ever to get the number of ratifications necessary for it to enter into legal force it would have to be weakened.

A series of local oil spill accidents in 1976-77 put the issue on the US domestic political agenda and led to threats of unilateral US action against foreign tankers in US waters unless a strengthened version of the MARPOL provisions on oil pollution were implemented quickly.[3] Accordingly in 1977 a further IMO conference confined the initial impact of MARPOL to oil pollution and strengthened its provisions in that area, but postponed entry into force of the remainder of the convention to a later date. With this modification the oil pollution provisions of MARPOL (which in effect amount simply to a substantial further strengthening of the original 1954 treaty) entered into force in 1983. The remainder did so only in 1987 – 14 years after the original negotiation. Even in 1994, fewer than 30 developing countries have ratified the convention.

The bulk of the international effort devoted to combating marine pollution following Stockholm, however, has gone not into supposedly global agreements like MARPOL and the London Dumping Convention but into regional agreements. This of course builds on a pattern which was already emerging by the time of Stockholm. We have already noted agreements among the Scandinavians and the North Sea states to cooperate to tackle oil spills, and how the London Dumping Convention itself sprang from an earlier regional agreement – the Oslo Convention – governing dumping in the North Sea and the North Atlantic. This pattern of the developed countries of the North charting the way was maintained through the agreement in 1974 on a convention (the Paris Convention) to limit contamination of the North Sea by land-based sources of pollution.[4]

The Paris Convention marked a major step forward from earlier marine pollution agreements. It was also a highly significant precedent for the big international actions to protect the atmosphere which lay in the future. This agreement, for the first time, shifted the focus of regulatory action from ma-

---

[3] R. Mitchell, 'International Oil Pollution of the Oceans' in P. Haas, R. Keohane and M. Levy, eds., *Institutions for the Earth* MIT Press, Cambridge MA 1993.

[4] P. Haas, 'Protecting the Baltic and North Seas', in Haas et al., eds, *Institutions for the Earth*.

rine pollution caused by a country's ships to marine pollution caused by its entire populace and industry. So, notionally at least, it controlled industrial and other discharges into all watercourses which would eventually find their way to the ocean. Its potential economic impact was thus far broader than that of earlier marine pollution agreements. It seems clear that the countries concerned were only able to arrive at a treaty of such potentially far-reaching effect by also including in it the stipulation that concrete antipollution measures established under the treaty required unanimous agreement. Indeed, there is no evidence that in the early years of its operation the Paris Convention imposed any controls on emissions that its parties were not already undertaking or willing to undertake for domestic reasons. Nevertheless it undoubtedly constituted a major new potential intrusion of international interest into the environmental consequences of domestic economic decision-making.

This potential became actual when, after a decade of rather slow progress in controlling land-based sources of pollution of the North Sea, the Germans, in frustration, convened the first of what was to become a series of meetings at ministerial level of the North Sea states (in 1984, 1987 and 1990). These occasions generated intense public and international interest in the condition of the sea, and thus political pressures on governments to do more to protect it. The result (which of course depended on the generally high level of environmental consciousness in the region, together with the responsiveness of democratically elected governments to public embarrassment) was that the North Sea governments found themselves agreeing with remarkable speed on significantly strengthened controls on emissions and dumping in the North Sea – occasionally going much further than they had intended (as with the unwilling agreement by the UK to accept controls on marine dumping in 1987). Indeed, the intensity of public interest in the conferences produced some decisions, such as an arbitrary 50% cut in certain pollutants, for which it is very difficult to find compelling economic or environmental arguments.[5]

5.3 Regional seas: the Mediterranean

The vast majority, at least in quantitative terms, of the regional marine activity which followed Stockholm stemmed from another child of the conference,

---

[5] Ibid.

UNEP. In 1975 UNEP decided that one of its 'concentration areas' was to be the ocean environment, and has since pursued its 'Regional Seas Programme' (since renamed Ocean and Coastal Areas Programme) to bring together regional groupings of states to tackle the problems of the local marine environment. This programme has contributed in large part to the boom in treaty numbers set out in Table 5.1. It produced nine regional agreements in the 1970s and six more in the 1980s – in other words, to about one-third of the total of all significant international environmental agreements negotiated in the two decades after Stockholm. The programme now has over 100 participating states and covers ten regional seas including the Red Sea, the Caribbean, the east Asian seas, and the south Pacific.

The scale of this programme, and the number of countries involved, is impressive evidence of UNEP's ability to pull regional groupings of countries together to discuss, and sign agreements on, their local marine pollution problems. Since most of the countries involved are developing countries the programme is also evidence of growing willingness in such countries at least to discuss the problem of marine pollution. It is a great deal less clear, however, how much the Regional Seas Programme has actually done to begin to reverse marine pollution, or to what extent it reveals a willingness on the part of the participants to make economic sacrifices to that end. With one exception, the programmes undertaken so far seem largely to be confined to generalized expressions of the need to tackle pollution, agreements to monitor the local marine environment and cooperate to tackle particular problems such as oil spills, and 'action plans' dealing with small local environmental projects – many of them financed by UNEP.[6] There is little evidence of the emergence of concrete regional programmes and standards intended to cut polluting discharges into the marine environment.

The exception is the action plan for the Mediterranean. This was the first of UNEP's Regional Seas Programmes, and in many ways has been an exemplar for the rest. It has gone further than the other programmes and, in so doing, has displayed in miniature many of the tensions and structures which

---

[6] UNEP, *Achievements and Planned Development of UNEPs Regional Seas Programme and Comparable Programmes Sponsored by other Parties*, UNEP, Nairobi 1982. See also L. Caldwell, *International Environment Policy, Emergence and Dimensions*, 2nd edn, Duke University Press, Durham, NC and London 1990, and H. French, *After the Earth Summit – The Future of Environmental Governance*, Worldwatch Institute, Washington DC 1992.

characterized the big environmental negotiations of the 1980s. It thus merits examination in some detail.[7]

The Mediterranean was one of the icons of the boom of environmental concern in the late 1960s. With the help of such publicists as Jacques Cousteau it was rapidly established in the Western public mind as an increasingly polluted body of water which posed a growing threat to health. The incidence of tar patches from oil spills, arguments about titanium dioxide dumping, and beach closures because of hepatitis outbreaks among tourists all contributed to this lurid picture. As elsewhere, however, concern about pollution was confined to the developed countries of the north shore of the Mediterranean. The developing countries in the South shared Northern concern about oil spillages (which were mostly caused by the North) and the potential loss of tourists, but otherwise were far more concerned to retain freedom of action to pursue economic development by whatever means they saw fit. President Boumedienne of Algeria said in the early 1970s that 'if improving the environment means less bread for Algerians then I am against it.' Thus a sequence of meetings of Mediterranean countries in the early 1970s, mostly convened by Italy or France to discuss the issue, produced nothing.

What these meetings revealed, however, was a high level of scientific uncertainty, not to say ignorance, about the precise nature and extent of the problems. The first proper scientific report on pollution of the Mediterranean was therefore prepared for the FAO (which was involved because of its interest in fish stocks) in 1972. It concluded that pollution, notably untreated sewage, oil and industrial waste, had 'reached a critical level'. The FAO thereupon set about drafting a treaty to deal with this; but in 1974, at French, Spanish and Italian instigation, it found itself shouldered aside (in an entirely characteristic example of turf competition between UN agencies) in favour of the newly created UNEP. Not only did UNEP bring wider environmental expertise to the task but it also had greater sympathy for developed country concerns – the FAO, for example, was not prepared to address land-based sources of pollution, given developing countries' fears of the implications regulation in this area might have for economic growth.

In view of the differences of approach between developed and developing countries, UNEP initially focused on improving scientific understanding. This

---

[7] What follows is heavily based on P. Haas, *Saving the Mediterranean*, Columbia University Press, New York 1990.

had two benefits. First, the construction of a scientific consensus about which types of pollution most needed tackling was a key step in persuading developing countries to help deal with such pollution at source. Second, the scientific effort itself fostered the transfer of research funds and technologies to the Mediterranean developing countries, so encouraging their continued participation. It also had the ironic effect of revealing that the initial alarm about the 'death' of the Mediterranean was exaggerated – as the head of UNEP later admitted – although this did not become clear at the time.

The upshot of this consensus-building period was agreement in 1975 and 1976 on a Mediterranean action plan and a cluster of legal agreements. None placed onerous environmental obligations on any of the parties (16 countries participated in the action plan and 12 in the legal agreements – and it is of some political significance that the countries concerned considered the matter to be of sufficient importance for Israel to be able to sit at the same table as a group of Arab states, and for Greece, Turkey and Cyprus all to be able to attend together). The action plan broadly encouraged countries to pursue projects of interest to them. The legal agreements contained general (but unspecific and unquantified) undertakings to control pollution from all sources, as well as specific protocols on dumping (modelled closely on the Oslo Convention) and on cooperation in the event of an oil or other environmental emergency. The negotiation of these agreements did reveal various particular tensions among the parties (such as French resistance to a Moroccan proposal for an emergency fund, and Italian opposition to a ban on sea dumping of titanium dioxide, since the Montedison company was doing so off Corsica) but it is arguable that the chief advance contained in them was not so much environmental as procedural. Their negotiation and conclusion built up a track record of communication and discussion on these issues among the countries concerned, and so helped establish mutual trust and a cadre of experienced officials necessary for the next stage.

That next stage, and the most difficult, was negotiation of the Land-based Sources Protocol, which took three years from 1977 to 1980. Like the Paris Convention (on which it was heavily based), this text shifted attention from pollution from ships to pollution from all land-based sources, marking a sharp expansion of the potential economic impact of environmental regulation. The developing countries were indeed deeply concerned about signing up to obligations which, by limiting their industrial emissions into the Mediterranean, could hinder their economic development. They were only edged into the negotiation at all on the basis of agreed scientific conclusions that 85% of the pollution in the Mediterranean came from land-based sources. They nevertheless held out firmly for a weaker system of standard-setting than the developed countries were demanding, and were partially successful in this. They also insisted that the protocol pay attention to 'the economic capacity of the parties and their need for development' and provide for a programme of technical assistance. For their part, the developed countries also had special interests to pursue. France and Italy resisted references to atmospheric pollution (for which their industry was mostly responsible) and France to radioactive pollution.

Despite these differences, a deal was struck and has since been ratified by the major Mediterranean states, North and South. This agreement goes a long way towards imposing the sort of controls that developed countries were seeking. It is strikingly detailed in its list of substances whose discharges are to be controlled, and includes a range of key industrial wastes. It leaves for future work, however, the precise levels of control to which these emissions will be subject. It was followed by the creation in 1979 of a Mediterranean trust fund to finance the Mediterranean plan activities, with the developed countries the major contributors, and (after the normal haggle about the site) the establishment of the headquarters and secretariat of the Mediterranean plan in Athens, where the participating governments meet annually to review progress. Since 1980 a great deal of work has been done, much of it financed by the European Community, to build on the understandings contained in the Land-based Sources Protocol; but that protocol remains the keystone of the efforts of Mediterranean countries to clean up the sea.

It is of course arguable that the Mediterranean plan in general, and the Land-based Sources Protocol in particular, is still in many of its aspects a framework which will only begin to bite when precise emission standards and other technical points are fully agreed – which is not expected to be until 1995. Only at that point will it become clear whether the countries concerned are willing to take on obligations of real economic significance. Even when the details are complete, it seems likely that the developing country parties (at least) will experience the same implementation problems with the protocol that we have seen them face with regard to their own domestic legislation. It is nevertheless striking testimony to the persuasive power of the scientific consensus, coupled with the growth of the habit of environmental cooperation among the Mediterranean nations, that they have managed to agree on so much. There is also good evidence that the process has given a boost to the evolution of domestic environmental policy, and the influence of domestic environmental ministries, in countries such as Egypt and Algeria – and that such states are beginning to take concrete steps to tackle their contribution to Mediterranean pollution. They are doing so, moreover, in the knowledge that, whatever they do, they can rely on continued action to cut pollution of the sea by the Northern Mediterranean states since the latter are under domestic environmental pressure and subject to EC and other commitments.

Finally, there is some tentative evidence that the Mediterranean plan constitutes a level three response to some forms of Mediterranean pollution – in other words, is having some effect on the state of the sea. The proportion of beaches judged unsafe for swimming has fallen from 33% in the mid-1970s to 20% in the mid-1980s. Much more sewage is properly treated and equipment installed to clean up ships' ballast. The feeling among regional scientists is that the level of pollution of the sea has been kept roughly constant despite fast industrial and population growth around its coasts.[8]

5.4 The Convention on International Trade in Endangered Species (CITES)

Apart from marine pollution, the other environmental area on which there was a track record of international environmental agreement prior to Stockholm was animal and plant conservation. This is an issue that has been driven much more by a Western public affection for wildlife than by any real assessment of the value (or costs) of its preservation. There had been popular lobbies in Western countries concerned about wildlife protection since well before the environmental boom of the 1960s; and there was a cluster of international organizations, both official and unofficial, concerned with the issue (notably the International Union for the Conservation of Nature – IUCN – and the World Wildlife Fund). As in the case of marine pollution, the Stockholm Conference, prompted by Western popular concern, helped carry work for-

---

[8] Haas, *Saving the Mediterranean*.

international developments

ward in this area. In particular it called for the negotiation of two treaties, the most important of which concerned trade in endangered species.

The idea of protecting wildlife by controlling trade in endangered species is not new. Paul Savasin, a Swiss conservationist, called for restrictions on trade in bird feathers for hats in 1911,[9] and the 1940 convention on nature protection and wildlife conservation in the western hemisphere (whose 15 signatories include 14 Central and South American developing countries) actually provides for restrictions on trade in a number of endangered species. But it was the huge growth in the trade in the in the 1950s and 1960s which provoked calls for global regulation. By the late 1960s up to 10 million crocodile skins were being exported each year, the US alone was importing 129,000 ocelot skins per year and Kenya's ivory exports reached 150 tonnes annually.[10] This led to calls by the IUCN from the early 1960s onwards for an international treaty to limit the trade. The issue was taken up by the US in 1969, close to the peak of popular environmental concern and following passage of the American Endangered Species Conservation Act. The Stockholm recommendation followed and in 1973 a plenipotentiary conference in Washington agreed on the text of the Convention on International Trade in Endangered Species (CITES) which entered into force in 1975 and now has over 100 ratifications.[11]

CITES is generally seen as the most successful of all international treaties to conserve wildlife. It bans trade in over 600 species and imposes controls on trade in over 26,000 others. The lists of protected species can be amended by a two-thirds vote at the biennial meetings of the conference of the parties. The treaty is interesting (and foreshadows the ozone layer agreements) in that it imposes a ban on trade with all non-parties unless they themselves introduce CITES-like controls. The speed with which the agreement was negotiated, and the number of ratifications it has received, result from the fact that it represents an attractive deal between the many developing countries who are concerned that their indigenous species (and often tourist potential) may be destroyed through poaching and illegal trading, and the developed coun-

---

[9] Lyster, *International Wildlife Law*, Grotius Publications, 1985.
[10] Ibid.
[11] G. Lukasser, 'Convention on International Trade in Endangered Species and the Impact of "Reputation" as a Compliance Device', Harvard Law School thesis 1991.

tries where popular pressure gives governments an interest in contributing to wildlife conservation.[12] The treaty was, moreover, amended in 1981 to strengthen the self-interest of parties in subscribing to it by promoting 'ranching', that is, allowing certain threatened species, or products from them, to be traded provided this was done in a way that ensured the conservation of the species concerned. It seems clear that this measure has contributed to the survival of a number of species, such as the Nile crocodile, by allowing them to be bred for use and trade, and is thus another example of environmental protection through financial self-interest.[13]

However, even given this widespread convergence of interest in making CITES work there are real, and revealing, implementation problems. First, despite the very high number of ratifications a number of key states are not party to the treaty – notably one or two, like South Korea, which act as centres for the animal trade (it has in fact been estimated that, despite the trade sanction referred to above, 30% of all wildlife transactions take place between parties and non-parties to CITES).[14] Second (and this was a safeguard demanded by a number of states during the treaty negotiation), the treaty itself permits parties to introduce 'reservations' excluding themselves from the provisions of the treaty with regard to particular species. Japan, for example, has maintained a number of such reservations, which it is abandoning only gradually under international criticism and occasional threats of US trade action. Third, there are the familiar implementation problems, and inability to control illegal trade, in a number of developing countries – coupled in some cases with an economically motivated unwillingness to apply CITES with any stringency. It took a major international campaign, for example, to get Thailand to pass the laws to implement CITES.[15]

---

[12] It also seems clear, though, that a number of developing country ratifications are principally the result of Western pressure. Thus Mexico adhered in 1990 following a campaign of Western lobbying and despite a Mexican official view that CITES was an 'aesthetic' convention that took a 'developed country approach' (personal communication). Animal protection is of course another area where the view in the West (where most animals are now seen as "wildlife") differs sharply from that in much of the South (where animals are sources of agricultural labour, much needed income, or pests).
[13] T. Swanson and E. Barbier, *Economics for the Wilds*, Earthscan, London 1992.
[14] Lukasser, 'Convention on International Trade in Endangered Species'.
[15] Ibid.

international developments

Finally, there are genuinely hard cases, of which the African elephant is probably the best known. Ivory poaching brought about a catastrophic fall in African elephant numbers over the 1980s (in Kenya, for example, the number fell from 65,000 to fewer than 20,000 in the decade), and growing demand that trade in ivory be banned under CITES. This demand was resisted by three groups: ivory traders such as Hong Kong, ivory consumers such as Japan and (less predictably) a small group of southern African nations such as Zimbabwe who were successfully conserving, and even expanding, their elephant herds, partly on the basis of earnings from the ivory of a few culled animals (using the 'ranching' provisions of the treaty). The 1989 CITES meeting imposed the ban over these objections, and maintained it in 1991, but the continued adherence of the southern Africans and others looks fragile unless some means can be found to meet their objections.[16]

Despite these various difficulties – which have helped maintain the world wildlife trade at a level of about $5 billion per year, of which one-third is estimated to be illegal[17] – there is good reason to believe that CITES too is a level three response to an international environmental problem, and has had a real impact on trade in endangered species. The ban on the ivory trade, for example, brought the price down sharply, so cutting elephant poaching by 80% (the number of poaching incidents in Kenya fell from 1,500 in 1989 to 30 in 1990).[18] Other species too, such as the South American vicuna, have recovered in numbers since being protected under CITES.

It has been observed that a key reason for the relative success of CITES is its administrative system, which at the time it was created was quite a new departure in environmental institution-building.[19] The establishment of a permanent secretariat, as well as the requirement on parties to set up bodies to enforce the convention, to communicate regularly with each other and the secretariat, and to meet regularly to review implementation of the conven-

---

[16] It has been persuasively argued (E. Barbier, B.Burgess, T. Swanson and D.Pearce, *Elephants, Economics and Ivory*, Earthscan, London 1990) that the only way the species can be preserved from extinction is by maximising its value to man, i.e. including culling herds for ivory.
[17] G. Porter and J. Brown, *Global Environmental Politics*.
[18] French, *After the Earth Summit*.
[19] Lyster, *International Wildlife Law*.

tion, all create regular opportunities for international, and public, exposure and pressure which in a number of cases have brought about visible changes of policy (for example, those of Japan and Thailand mentioned above). It also seems likely that the fact that CITES is enforced through national customs authorities, which tend to be relatively well-developed government agencies in developing countries, has helped. Thus 'there is no chance of CITES becoming a "sleeping treaty" which its parties can safely ignore'.[20] This, of course, is a lesson we have already seen in connection with the North Sea agreements.

5.5 Atmospheric pollution: acid rain

If marine pollution and wildlife conservation were well-established areas of international business by the time of Stockholm, atmospheric pollution certainly was not. Action to tackle atmospheric pollution at the national level of course preceded Stockholm. The UK took stringent legislative action to deal with London's 'killer smogs' in 1956, and the US passed its Clean Air Act in 1970. However, taking the Stockholm conclusions as reasonably representative of the environmental agenda as it was seen in 1972, it is striking in view of subsequent events how little reference they contain to the international aspects of atmospheric pollution. The atmosphere is of course duly identified as one of the six media which should be safeguarded, and there are three prescient recommendations touching on the need to watch for the effect of human activities on the climate (which both stemmed from rising scientific interest in the subject at the time and helped to give that interest further impetus); but Stockholm is otherwise eloquently silent on the problems of the international atmosphere, especially in comparison with the pages of recommendations devoted to marine pollution, species conservation or even fishing.

Thus air pollution was not seen by most governments as an international problem, either at Stockholm or for some years subsequently. The only exceptions to this general attitude were the governments of Norway and Sweden which had become alarmed by Swedish scientific evidence that other countries' emissions of sulphur dioxide and nitrogen oxides, mostly from the

---

[20] Ibid.

chimneys of large electricity-generating plants, were coming down in their countries as acid rain. That is to say, the emissions were dissolving in the water vapour of the atmosphere, often being carried hundreds of miles by the winds, and making rainfall in places remote from their place of origin (notably in Scandinavia) significantly more acidic – with damaging effects on plant life, fish stocks and even some buildings. Indeed, one reason why Sweden offered to host the Stockholm Conference was to draw international attention to this problem. But the efforts of the two governments to push this case at Stockholm were met with scepticism. Only in 1977 did an OECD study begin to persuade governments that long-range transmission of sulphur dioxide and nitrogen oxides actually took place, and that acid rain in certain Scandinavian countries was indeed largely the consequence of emissions elsewhere. This finding prompted the US and Canadian governments to undertake a joint study which revealed that 50% of the acid rain in Canada resulted from US emissions.[21]

In the US-Canada case the upshot of this finding (despite a long history of good environmental cooperation between the two countries, notably over cleaning up the Great Lakes) was more than a decade of ineffective Canadian pressure on the US to cut its emissions, with the US consistently arguing that there was insufficient scientific evidence that its emissions were doing any real damage. Then, in 1990, in response to domestic popular concern about air pollution, the US amended its clean air acts to cut sulphur dioxide emissions by 35%, which also enabled it to sign an emissions reduction agreement with Canada (although there is no sign that Canadian pressure brought about deeper cuts than those fixed on for domestic reasons).

In the European case, rising concern about acid rain coincided fortuitously with a moment in the East-West detente process when the East was looking for an item of multilateral business to do with the West other than in the areas of human rights and arms control. Thus a negotiation was launched among all European countries (and the US and Canada) which led to the 1979 Convention on Long Range Transboundary Air Pollution (LRTAP). It was not an easy negotiation. The Scandinavians, who were recipients of acid rain, demanded cuts in emissions of sulphur dioxide and nitrogen oxides, while the

---

[21] M.E. Wilcher, *The Politics of Acid Rain*, Avebury, 1989.

major exporters of such pollution – notably the UK, France and West Germany – resisted such cuts, as the US had done, on the grounds that there was insufficient scientific evidence that their emissions were doing any damage. No doubt they were also concerned about the costs. In the case of the UK, for example, a substantial programme of sulphur dioxide reductions would cost billions of pounds, and would bring little national environmental benefit as most of the emissions were removed by the prevailing winds. Eventually the Scandinavians gave way: as a result, LRTAP went no further than general statements of good intent and setting up machinery for information collection and exchange.

The convention also provides, however, for regular meetings between the parties, and anticipates that there will be subsequent protocols (additional, more specific, treaties flowing from the main treaty). These provisions, together with accumulating scientific evidence of the damage done by acid rain, have proved effective mechanisms for ratcheting up the original commitments. In particular, they have generated regular occasions for public and international pressure in much the same way as with the North Sea and CITES agreements. Indeed, by the time the parties to the convention first met subsequently in 1983 a number of them had changed position on cuts in sulphur dioxide emissions – France because its nuclear energy programme meant that its emissions were in any case falling, and Germany because new evidence that acid rain caused forest death had created powerful domestic political pressures for reductions. Most East European countries were willing to agree on sulphur dioxide cuts in the interests of substantively building up their relations with the West. As a result the '30% Club' came into existence. This grouping of countries willing to make a 30% cut in sulphur dioxide emissions began placing heavy pressure on other governments to take on the same commitment through a sequence of highly public international meetings over the next two years. One by one, countries succumbed to a combination of external diplomatic and internal environmentalist pressures. Finally, in 1985, the large majority of the signatories of LRTAP agreed on the 'sulphur protocol' embodying the 30% commitment. The major non-signers were the UK, Poland and Spain. This protocol was followed in 1988 by a similarly hard and publicly fought nitrogen oxides protocol in which parties agreed to freeze their nitrogen oxide emissions (with a number pledging themselves to go further). The same year also saw final agreement on a European Community

international developments 107

directive on the subject (the Large Combustion Plants Directive) in which, after five years' negotiations, the UK and Spain both agreed to substantial reductions in their sulphur dioxide emissions.

This cluster of agreements is quite heartening evidence of the worth of environmental diplomacy, at least as conducted among industrialized countries. It has undoubtedly helped to maintain downward pressure on sulphur dioxide and nitrogen oxide emissions in the European region, and this is beginning to affect the acidity of the rainfall.[22] While many of the countries concerned were undoubtedly pushed into reducing their emissions, by domestic factors it is clear that in a number of cases the international process, and the popular interest and pressure that it created, played a large part in bringing about a change of policy which (for example, in the case of the UK) entailed substantial domestic costs for environmental benefits which would mostly accrue to other countries. It is nevertheless important to note that these negotiations took place among a group of developed countries, with a multiplicity of cultural, economic and political links and highly developed public opinion on environmental issues. Even so, it took a decade of intense diplomacy to produce effective international action on acid rain in Europe.

5.6 Environment in the European Community

If atmospheric pollution has been one recent growth area for international environmental agreements, the other has been the increasing acceptance of the environment as a whole (as opposed to specific areas, like marine pollution) as a subject for international cooperation. Early examples of regional framework agreements which have included the environment (or aspects of it) as a subject for cooperation include the 1974 Nordic Convention on Protection of the Environment and the 1978 Amazon Pact Treaty. But undoubtedly the most significant and most successful example of regional cooperation across the board on environmental matters is provided by the European Community. Indeed, the history of Community environmental action and politics contains many features subsequently reproduced on the global scene.[23]

---

[22] M. Levy, 'European Acid Rain, the Power of Tote Board Diplomacy', in Haas et al., eds, *Institutions for the Earth*.
[23] For background see e.g. A. Liberatore, 'Problems of Transnational Policymaking – Environmental Policy in the European Community', *European Journal of Political Research 19*, 1991.

The Community was established in 1957, well before the environment as such was seen as a subject for public policy. The environment is not mentioned in the Community's founding document, the Treaty of Rome. However, a central aim of the Community is to unify the economies of its member states and, with the advent of environmentalism in the 1960s, it was rapidly appreciated that economic unification was bound to have an environmental dimension since different environmental standards in different member states would create distortions of competition and obstacles to trade. Moreover, the transboundary nature of many environmental problems (marine and river pollution, for example) makes it natural for the relatively compact and closely linked nations of the community to cooperate to deal with them. Thus the European Parliament first began to address pollution problems in 1967. In 1967, too, the Community passed its first environmental directive (i.e. law), which neatly combined the Community's trade vocation with its new-found environmental concern by regulating the classification, packaging and labelling of dangerous substances.

> The Environmental Action Plans ... constitute a sort of political portmanteau into which the parties can bundle their current environmental concerns in a suitably non-binding way ... their development shows interest shifts of emphasis ... But the real backbone to community to environment policy was the development of Community environmental law ... the Community institutions in Brussels provide for more or less perpetual negotiation, often under intense press scrutiny.

The political sanctification of Community environmental activity more or less coincided with the Stockholm Conference, and generated a similar product. In October 1972 a summit meeting of Community heads of state and government for the first time addressed the issue of environmental protection and called for a 'community environmental action plan'; this was duly approved a year later. This plan (and its four successors, in 1978, 1982, 1987 and 1992) resembles the Stockholm conclusions in that it is a level one document. It has no legal force and constitutes a sort of political portmanteau into which the parties can bundle their current environmental concerns in a suitably non-binding way. To the extent that those concerns also reflect the legislative intentions of the governments involved there is of course (as in the case of Stockholm) some overlap between the contents of each plan and subsequent legislative activity. But there are also large areas of difference. Overall, it is very hard to be confident that the plans have had a significant impact on subsequent action in the Community or in the member states. Nor did they mark in any sense a shift to the Community of exclusive rights of legislation on environmental issues. Individual member states continued, and continue, to legislate nationally on matters of concern to them.

Thus the environmental action plans should be viewed more as a political signal of areas of shared environmental concern in Europe than as concrete legislative programmes. Viewed in this light, their development over the 20 year history of Community environmental activity shows interesting shifts of emphasis (which also took place elsewhere in the developed world) from curing environmental problems to preventing them, from viewing the environment as a separate policy area to viewing it as a necessary component of all government policy-making, and from tackling problems through regulation to tackling them through other (notably economic) instruments.

The real backbone of Community environment policy, then, was not the action plans but the development of Community environmental law. This has taken place on a large scale. The Community now has over 280 items of environmental legislation ranging across such key issues as water quality, waste management, air pollution, control of dangerous chemicals, wildlife protection and noise control.[24] In general the pace of legislation has accelerated (though there was a slowdown in the second half of the 1980s), with 79 items adopted in the 1970s, 109 in the 1980s and 35 in the two years 1990-91. Many of these directives are narrowly technical, but a significant number constitute major acts of regional environmental cooperation which in any other geographical context would have been seen as landmark environmental treaties. Random examples include a Community-wide system for the approval of hazardous chemicals, Community-wide requirements for the assessment of the environmental impact of major industrial or other development projects, and Community-wide standards for motor vehicle exhaust emissions.

---

[24] For more details see N. Haigh, *EEC Environment Policy*, Longman, Harlow 1990.

This enormous number of agreements is not the product of any abandonment by member state governments of concern for their own national interests in favour of some abstract concern for the purity of the European environment. It is rather, the outcome of intense and extended negotiations in which member states find themselves under a number of pressures to adjust their national priorities to respond to the concerns of their Community partners. The first such pressure is environmental. Because of the geographical compactness of the Community, a pollution problem in one member state tends to be shared by others and so to require a Community-level response. There is no point in the Netherlands alone reducing chemical discharges into the Rhine. Second, there is the effect of domestic popular opinion – which often finds it difficult to understand why its government is holding out against a Community agreement which will have the effect of improving the environment. Third, there is the economic interconnectedness of the Community which makes it crucial, for example, that the member states agree on common environmental standards for widely traded products (such as automobiles). Another effect of the close economic links is that Community governments which wish to raise domestic environmental standards tend to press for the introduction of those higher standards at Community level so as to limit the competitive consequences for its own industry (the 'level playing field' effect). Fourth, there is some need for each Community member state to accommodate the political priorities of other Community governments, given the very wide range of business they do together. While it is rare for any government to make an explicit link between progress in an environmental negotiation and progress in another area of Community activity, governments do nevertheless tend to fall in with an emerging consensus in one area of negotiation in order to foster the emergence of such consensuses in other areas which may be of more concern to them. This interplay of national and European pressures and interests is given full scope through the Community institutions in Brussels which provide for more or less perpetual negotiation, often under intense press scrutiny.

An illustration of the way the process operates is provided by the Large Combustion Plants Directive referred to above. This was originally proposed in 1983, largely at the behest of a German member of the European Commission (the executive arm of the European Community) at a time when Germany had just become concerned about the problem of acid rain because of the effect it was having on German forests. The original aim was that all members of the Community should agree to become members of the 30% Club, but this was rapidly abandoned in the face of the determination of the relatively underdeveloped Ireland and Greece to expand industrially, and emit more sulphur dioxide in the process. The proposed emissions targets were adjusted to accommodate this and bring intense pressure to bear on the UK which, as we have seen, was at that time the only major Community emitter not to have taken on the 30% commitment. The UK held firm, arguing (as in the case of the US dealing with Canadian demands for emission reductions) that more scientific evidence was needed to justify the restriction. At the start of 1986 the position was further complicated by the accession to the Community of Spain and Portugal, both of them southern countries with ambitions for fast economic growth and the former a major sulphur dioxide emitter. Again the targets were adjusted. The negotiations nevertheless remained deadlocked until an intense burst of work by the German chairman of the Community's Council of Environment Ministers in mid-1988 resolved the remaining disagreements. Spain agreed to freeze its emissions by 1993 and cut them subsequently, provided it was exempted from another Community requirement on non-polluting technology. The UK, under political pressure because of the renewed popular upsurge of environmentalism in 1988 and under economic pressure because of the impending privatization of its electricity industry, finally agreed to cuts slightly short of what was needed for it to join the 30% Club.[25] The negotiation had taken five years. In the course of it the original proposal had been radically modified to accommodate particular interests of particular member states. Constant public and political pressure, as well as a change in domestic circumstances, were required finally to bring everyone on board. Nevertheless, the outcome undoubtedly marks a shift in a number of member states (notably Spain and the UK) to do more to curb sulphur dioxide emissions than they originally intended, and as such constituted a major success for European environmental cooperation.

---

[25] M. Levy, 'The Greening of the United Kingdom: An Assessment of Competing Explanations', paper delivered at the August 1991 meeting of the American Political Science Association, Washington DC.

By the mid-1980s, then, the Community had become an experienced manufacturer of environmental legislation, to the point where Community legislation was becoming the major source of new environmental law in a number of member states. By now two problems were apparent. The first was implementation. In the European Community, as in certain developing countries, the passage of an environmental law is not automatically followed by any change of circumstances on the ground. By no means all of the impressive level two Community response was translated into level three. A number of member states were simply failing to transpose Community requirements into their domestic legislation or enforce them. This problem (which applies across the whole corpus of Community law) was particularly acute in the environmental field because of the speed with which legislation had been enacted and because the very visibility of environmental issues led to a high volume of public complaints – to which the European Commission, which is responsible for ensuring that member states implement Community law, has to respond. In 1990 the Commission had nearly 400 outstanding cases (mostly resulting from public complaints) of alleged non-implementation of Community law by member states. As a result it has in recent years had to put significantly more resources into environmental enforcement.

The second problem was the lack of any reference to the environment in the basic treaties of the Community. The Community is a highly legalistic organization. It is therefore striking evidence of the need for a shared corpus of environmental law that, despite the lack of any explicit treaty basis, Community law on the environment accumulated so rapidly. The lacuna nevertheless grew more and more glaring. As the environment rose once again to political prominence throughout the West in the mid-1980s it looked increasingly anomalous that there was no mention of it in the basic documents of the major organ of European political and economic cooperation. The omission also had practical disadvantages. It meant, for example, that environmental considerations had no standing in relation to other areas of Community policy, and the Community's fast-growing 'structural' spending on internal economic development was no less prone than other multilateral development spending at this time to ride roughshod over environmental concerns.

The gap was finally filled in 1987 when the Community treaties were modified by the Single European Act (SEA). The main aim of this measure was the creation of the 'Single European Market' but, partly at the insistence of member states like Denmark which saw a close link between economic integration and shared environmental standards, it also included a chapter on the environment. This text was a hard-fought compromise between those (mostly European integrationists) who wished to see the Community role in European environmental policymaking grow still larger (and so demanded, for example, adoption of European environmental legislation by majority vote) and other member states which were more conscious of the potential costs and loss of autonomy in transferring more environmental authority to the Community. The outcome was an agreement which broke new ground in a number of ways. It explicitly made environmental protection a subject for Community action. It did not allow decision by majority vote for purely environmental issues, but did allow it for issues of, for example, common standards (including environmental standards) which will facilitate the free flow of goods. Moreover, for the first time in any major international treaty the SEA introduces the principles that environmental action should be preventive, that environmental concern should be a component of other Community policies, that there should be a 'high standard of environmental protection', and the 'polluter pays' principle (the principle under which agents should be financially responsible for the environmental consequences of their activities).

The Single European Act was rapidly superseded.[26] In 1990 the Community (again for non-environmental reasons) launched another round of negotiation on the modification of the basic treaties. In the environmental context two things had changed crucially since negotiation of the SEA. First, the world was by now experiencing another boom of environmental concern; second, the Southern states of the Community, particularly Spain, were (in a manner very reminiscent of the developing countries at Stockholm) much more conscious of the potential constraint on their economic growth which might result from tough Community environmental standards, as well as the costs to them of the imposition of those standards on Community aid projects within their borders. At the time there was some talk of a Community 'North-South gap' on the environment, with the South arguing that Community policy paid excessive attention to Northern problems, such as industrial pollution,

---

[26] C. Kim, 'The Making of Maastricht: Environmental Policy', paper presented to the Harvard International Environmental Institutions Research Seminar, October 1992.

rather than Southern problems such as lack of environmental infrastructure and water shortages. It is certainly the case that up until this time Community policy had been almost exclusively devoted to environmental *regulation* as opposed to the financing of environmental projects (indeed, up until 1986 environmental spending accounted for less than 0.1% of the Community budget). As a result, Spain in particular made it quite clear that there could be no extension of majority voting on the environment unless there was also some arrangement to help the Southern states with their problems. The upshot was an agreement which indeed extended majority voting but also established a 'cohesion fund' to help Southern states with, *inter alia*, environmental infrastructure. In addition it introduced for the first time into an international treaty the 'precautionary principle' under which the lack of absolute scientific certainty should not justify inaction where environmental damage seemed probable.

One other fast-growing area of Community environmental activity must be mentioned briefly. With the growth of its corpus of shared environmental legislation the Community has also increasingly acted as a unit in wider environmental negotiations. To some extent this development stems from the Community treaties, which require the Community member states to take a common negotiating position on certain named issues (such as trade) and those where the Community has internal legislation. Thus the Community has negotiated as a unit on environmental issues with trade aspects (such as CITES and the ozone layer negotiations). On other issues the simple habit of Community environmental cooperation, the broadly shared interests of European countries, and the constant meetings of European environment ministers, where differences can be ironed out and common positions assembled, have often enabled Community member states to act together, and make a significant impact by so doing. It also seems likely that the habit among the Twelve of ironing out their own differences first and trying to speak with one voice has often simplified and accelerated the wider negotiation (though the process is undoubtedly frustrating for outsiders and has led to claims that the Community tends to adopt 'lowest common denominator' positions and so has slowed certain negotiations down).[27] This is not to say that member states agree on all issues in all international negotiations. They certainly do not, as we will

---

[27] See e.g. P. Haas, 'Protecting the Baltic and North Seas', on the North Sea negotiations, and R. Benedick, *Ozone Diplomacy*, Harvard University Press, Cambridge MA, 1991, on the ozone layer negotiation.

international developments

see in particular examples below; but the instinct to look for a common Community position before proceeding to any wider forum is now very strong.

Overall, the growth of both internal and external Community environmental cooperation in the two decades following Stockholm is a quite remarkable phenomenon. It has of course depended crucially on the very special political, economic and institutional links that the Community member states have evolved, and continue to evolve, among themselves. But it is also a clear testimony to the ability of that group of states, despite the very real differences of environmental situation and interests among them, to find cooperative solutions to common environmental problems. It is worth noting again the factors which have been crucial to this result – intimate economic and geographic linkages, a broadly shared level of economic development and political culture (including in particular democratic systems and relatively high levels of popular concern about the environment) and a more or less perpetual process of negotiation which maintains both political and public pressure on governments to move towards consensus.

5.7 Nuclear issues

One area of environmental business where, despite intense public interest, very little was done in this period at the international level was the issue of nuclear safety. This has probably been the most conspicuous example of an environmental issue where an extreme polarization of views has made debate sterile at both national and international levels. At the national level, throughout the West, the pro-nuclear lobby (which has tended to include governments, energy ministries and their supplying industries) has taken its stand on the science (indeed, one unusual feature of the nuclear issue is that the environmental movement here is arguing *against* most of the scientists). The view of most qualified scientists is reasonably clear. The radiation risks associated with nuclear generation of electricity can, with proper safeguards, be made acceptably low by comparison with other forms of power generation, and there are safe technological solutions to the other problems posed by nuclear power – notably the disposal of radioactive waste.[28] The nuclear industry has

---

[28] See e.g. G. Foley, *The Energy Question*, 4th ed, Penguin, London 1992; H. Lewis, *Technological Risk*, Norton, New York 1990.

caused fewer deaths per unit of power produced than other forms of power generation. Agreed radioactivity standards are very tight: exposure standards in the industry, for example, amount to less radiation than one could receive from a colour television set, and radioactivity from low-level waste is less than from a luminous wristwatch.[29] This message has not, however, been well put over. Repeatedly the nuclear lobby has seemed secretive and dismissive of public fears on the issue.

We have seen that hostility to nuclear power became at a very early stage an important feature of the environmentalist view of the world. Environmentalists have tended to emphasize the uncertainties attached to a technology as new and complex as that of nuclear reactors, and the damage that could be done if things were to go wrong. They have questioned the science,[30] and have placed strong emphasis on incidents such as that at Three Mile Island in Pennsylvania in 1979 (where, although there was nearly a serious accident, no-one in fact was hurt). They have been able to draw on uninformed public fears of radioactivity (which, as repeated surveys have shown, causes more public alarm than statistically much greater hazards such as driving, smoking and coalmining),[31] which have led to the 'not in my backyard' type of local protests against any proposal for a nuclear installation.[32] Thus nuclear development throughout the 1970s and 1980s saw a sequence of often violent confrontations between the two camps throughout Western Europe and the US, notably at Whyl (Germany), Crays-Malville (France), Kolkar (Netherlands) and Seabrook (New Hampshire, US).

This alignment of forces has produced different outcomes in different countries. France has been the standard-bearer for the nuclear generation of electricity. After weathering an early round of public protest, and aided by a

---

[29] Lewis, *Technological Risk*.

[30] Thus the scientific assessment of the risk attached to the burial of nuclear waste is 'as negligible as it is possible to imagine... at least a factor of a million lower than (other common technological risks)' (ibid.), while even so relatively objective a source as Tolba et al. (The World Environment) refers to 'serious environmental hazards which will probably become more serious as time goes on'.

[31] A. Blowers, D. Lowry and B. Solomon, *The International Politics of Nuclear Waste*, Macmillan, London 1991. For a good exposition of the antinuclear case see also Christopher Flavin, *Reassessing Nuclear Power: The Fallout from Chernobyl*, Worldwatch, Washington DC, 1987.

[32] Foley, *The Energy Question*.

international developments

tradition of technocratic government and a lack of alternative indigenous energy sources, the French government has pressed ahead with a nuclear programme which now supplies 75% of the country's electricity (compared with an average of 23% in the OECD as a whole). Moreover, public opinion polls regularly show significantly more support for nuclear power in France than in other West European countries. In east Asia too, notably Japan, Korea and Taiwan (countries which share with France a tradition of technocratic planning in their style of government), the nuclear sector is still being expanded. But in other Western countries, including the UK and the US, as well as the main countries of Eastern Europe (where of course public opposition was not until recently a problem),[33] the nuclear industry, after a dynamic start, has consistently had to scale back its projections of growth. Indeed, there are now no plans for nuclear expansion anywhere in Europe other than France. There has been no nuclear reactor order in the US for 20 years which has not subsequently been cancelled. Sweden decided by referendum in 1980 to phase out its hitherto ambitious nuclear programme. There are intense political pressures, notably in Eastern Europe, for the closure of older reactors which do not meet modern standards. The environmental movement has not singlehandedly brought about this result. The high costs of nuclear electricity by comparison with the alternatives, and the scaling back of earlier overoptimistic projections of Western demand for electricity, have played a major role. But there is no doubt that widespread public doubts about nuclear safety,[34] encouraged by the hostility of the environmental NGOs, have played a significant part.[35]

A subject so highly polarized at the domestic level was hardly likely to be profitable matter for international discussion, and so it has proved. For a long time, and despite the evident transborder implications of the issue, there was effectively no international discussion of nuclear safety or waste disposal issues. The original intention of the European Community to pool its ownership of nuclear material (agreed in the 1957 Treaty of Rome) was never

---

[33] Although see Pryde, *Environmental Management in the Soviet Union*, for hints of early public doubts even there.
[34] Even before Chernobyl public opposition to new nuclear facilities stood at 67% in the US and 37% in the UK, and was undoubtedly higher in the vicinity of any proposed nuclear plant. See Tolba et al., *The World Environment*.
[35] Foley, *The Energy Question*.

implemented. Nuclear issues were not discussed at Stockholm (although, as we have seen, that conference did give rise to a major row between signatories and non-signatories of the Test Ban Treaty, whose product was a strikingly bland reference to radioactive pollution in the conference conclusions). The developed countries deliberately kept the subject off the agenda of the 1981 UN conference on new and renewable sources of energy.[36] When the subject did find its way on to international agendas, this tended to take the form of countries with no nuclear industry (such as Ireland) trying to use the international system to impose tougher standards on neighbouring countries (such as the UK). In the same vein the mid-1980s saw pressures from certain non-nuclear European Community states to introduce European Community standards for nuclear safety and a Community inspection system. These were firmly seen off by the nuclear states of the Community, notably, of course, France. At the global level, occasional suggestions that the International Atomic Energy Agency (whose principal responsibility is nuclear non-proliferation) should interest itself in nuclear safety issues were similarly squashed. The meetings of parties to the London Dumping Convention, which bans the dumping of high-level nuclear waste at sea, saw years of repetitive argument about the possible extension of this to low-level nuclear waste. These were concluded by a hotly contested vote in 1983 in favour of a (non-compulsory) moratorium which, since it could not be justified on scientific grounds, was lamely based on economic and social factors.[37] For a long time the only real international work on the issue tended to be highly technical – such as agreements in 1962 and 1971 on liability in the event of a radiation accident at sea. We will see below how it took a major accident to change this pattern.

5.8 Development of the north-south agenda

Nuclear issues were not the only ones that did not make much progress in the period under discussion. The period is punctuated by a string of (mostly UN-sponsored) conferences, negotiations and reports which may (like Stockholm)

---

[36] Caldwell, *Between Two Worlds*.
[37] S. Boehmer Christiansen, 'The Role of Science in the International Regulation of Pollution', in Andresen and Ostrong eds, *International Resource Management*, Belhaven 1989.

have done something to raise awareness of particular problems within particular national governments but have otherwise left little mark. Many of these encompass both specific issues and broader aspects of the North-South relationship. These developments – or lack of developments – are of great importance in understanding the background to UNCED.

Two major sets of global negotiations with significant implications for the environment came to grief between Stockholm and Rio. The first was the third UN Conference on the Law of the Sea (UNCLOS), which lasted nine years from 1973 to 1982. This was intended to tackle the full range of marine law issues, including the growing proportion of them touching upon the environment. With regard to fish conservation, for example, the increasingly apparent problem of depletion of ocean stocks through overfishing was tackled by 'enclosing' 90% of those stocks through the creation of exclusive economic zones, 200 miles off nations' coasts (though there is evidence that even this drastic step is not working).[38] Marine pollution, as we have seen, was already the subject of a growing range of special treaties, so UNCLOS confined itself to imposing a (more symbolic than effective) general duty to conserve the marine environment from pollution from all sources, and revising and clarifying the attribution of legal rights and responsibilities (notably for enforcement) among flag and coastal states. But the rock upon which the negotiation foundered was the mineral wealth supposedly available on the beds of the oceans. In the draft of the treaty these reserves were declared the 'common heritage of mankind' and complex machinery was created to ensure that developing countries should share in the benefits of their exploitation. Following changes of government in the US and the UK these arrangements were seen as a significant disincentive to private enterprise to develop the technology to exploit the ocean bed. It was thus not until 1994 that the treaty was ratified by the 60 countries necessary for it to enter into force and it looks highly unlikely that the 'common heritage' provisions will ever operate in the way originally intended.

The other failed negotiation was that on the so-called 'New International Economic Order' (NIEO).[39] This was not centrally an environmental nego-

---

[38] M. Peterson, 'International Fisheries Management', in Haas, Keohane and Levy, eds, *Institutions for the Earth*.
[39] Cf. Hart, *The New International Economic Order*, Macmillan, London 1983.

tiation but a developmental one with significant environmental overtones. It figures here because it was the most important international outing between the Stockholm Conference of 1972 and the Rio Summit of 1992 for the 'North-South environmental bargain' argument – that the North should in its own interests give more economic help to the South so as to help tackle the environmental threat that it itself otherwise faced from that quarter. The background lay in demands by developing countries through the 1960s for a redistribution, in their favour, of international wealth and economic power. We have already seen how at Stockholm these demands became linked to the question of the global environment but made little progress towards fulfilment. The Southern countries continued to receive rather short shrift from the North until they were given a substantial boost by OPEC's use of the 'oil weapon' in 1973. The lesson drawn by many developing, and some developed, countries from the quadrupling of oil prices in that year was that the developing countries might be able to use their natural resources as an instrument to tilt world economic arrangements more their way. The upshot was eight years of negotiation, including two special sessions of the UN General Assembly, in which bland general formulae were used to conceal the size of the gap between the demands of the South (for such things as debt relief, assured prices for Southern commodities and more Western aid and technology) and the unwillingness of the North to concede them.

At this time the Club of Rome produced a report arguing that it was important for the North, not least for environmental reasons, to do more to help the South. Two other 'wise mens' reports' of the era (the Brandt Report, *North/South: a Programme for Survival*, and the *Global 2000 Report to the President of the United States*) took much the same line.[40] The argument, basically, was that if Southern poverty was not tackled then population growth and resource depletion there would eventually undermine prosperity in the North as well. There were some Northern concessions to this line of argument. In many ways the most persuasive and successful of these have been the four successive Lomé Conventions (1975-89) which have linked the Eu-

---

[40] Brandt Commission, *North/South: A Programme for Survival*, MIT Press, Cambridge MA, 1980; US Council of Environmental Quality and Department of State, *Global 2000 Report to the President of the United States*, Washington, DC 1980.

ropean Community to 69 developing countries in Africa, the Caribbean and the Pacific. In these agreements the Community has committed itself to assured quantities of aid and guaranteed market access for certain products, while the developing country participants have guaranteed Community access to certain of their raw materials and markets as well as making undertakings on such issues as human rights, population, the role of the private sector and (increasingly) environmental protection.

But Lomé was the exception. At no point in the NIEO negotiation proper did the major Western countries see themselves as sufficiently threatened by the economic and environmental circumstances of the South to consider making the sort of sacrifices that were demanded of them. The whole process finally collapsed at the 1981 Cancún Summit where the US, UK and Germany rejected Southern calls for global negotiations. We will see below that a decade later many of the same arguments were dusted off for use in Rio in 1992.

There were several more specific negotiations with a strong North-South component that also made relatively little headway. An early post-Stockholm example is provided by the UNEP-inspired UN Conference on Desertification (UNCOD) in Nairobi in 1977. This conference, with 94 participant nations, was promoted principally by developing countries threatened by desertification. It produced a 'plan of action' to begin to tackle the problem, but left funding voluntary. As a result, and given that the developed countries themselves did not feel much threatened by desertification, little was forthcoming. A follow-up conference seven years later noted that virtually nothing had been done and that the problem had in fact grown worse. Similarly, the UN Conference on New and Renewable Sources of Energy (Nairobi, 1981) agreed on an action plan but no machinery for implementation (the developed countries resisted the establishment of a new UN agency). In 1982, the tenth anniversary of Stockholm, an 'anniversary meeting' of 70 ministers and heads of government was held in Nairobi; it produced little beyond a rather downbeat assessment of the state of implementation of the Stockholm recommendations, coupled with a lot of ringing speeches about the inequity of the international order and the evil of nuclear weapons.[41] The one concrete product of this meeting was, as the result of a quite unnecessary row between Japan and

---

[41] *The Annual Register – a Record of World Events*, 1982, Longman, Harlow, 1993.

Sweden, to postpone by a year the establishment of what was to become the Brundtland Commission (see below). In the same year the UN General Assembly adopted, as the result of a developing country initiative, the 'World Charter for Nature' which sets out 24 mandatorily phrased conservation duties for nations and individuals including such sentiments as 'the population levels of all life forms, wild and domesticated, must be at least sufficient for their survival.' It has had no visible subsequent impact.

Not all global environmental activity in the years following Stockholm has been as unsuccessful as these last examples suggest. We have already seen that despite a sequence of bruising international conferences on the subject, low-key Western aid for population projects has contributed materially to cutting the global rate of population growth. Another example of a similar process at work is offered by the World Conservation Strategy which was launched in 1980, after three years' intense work, by the IUCN with the assistance of UNEP. It has been described as 'the most thorough stocktaking of the earth's natural resources since Noah's Ark'. It displays many of the symptoms of a number of the failed initiatives discussed above. It argues that humanity's survival is threatened by the abuse of nature, and that only 20 years remains in which to limit the damage. It then sets an ambitious programme of national and international objectives and targets for species management and conservation if this is to be done – with no visible machinery for implementation or funding. Nevertheless the IUCN has quietly but effectively pursued the objectives of the strategy with a large number of governments, in particular through trying to show them their own economic interest in conservation activity (for example through expanded tourism) to the point where 45 countries have now developed national conservation plans. As we will see below, this does not mean that the world's biodiversity problems are by any means solved, but it does constitute a significant international effort to begin to reverse the tide.

5.9 Conclusions

This swift survey of international environmental activity in the years following Stockholm suggests a number of patterns. The first is the striking speed and success with which developed countries have learnt to cooperate to tackle

international developments

shared environmental problems. We have now seen a number of cases where that cooperation has gone beyond 'cosmetic' treaty-signing to satisfy some domestic lobby to the point where nations are making real adjustments to their domestic environmental policies in response to the pressures generated by the international negotiation process. The international community has rather rapidly found ways of achieving level three responses to at least some international environmental problems. This has been true of the acid rain saga, the North Sea ministerial meetings and regularly within the European Community. To a large extent this development reflects the susceptibility of democratically elected governments to pressure by environmentally aroused electorates as well as industrial demands for a 'level playing field'. It is also the product of a particular style of environmental treaty in which a basic framework agreement is supplemented by a programme of continuing work on concrete obligations, often entailing high-level meetings which attract public attention and pressure.

Groupings involving developing countries, whether global or regional, have so far shown less capacity to bring about changes of individual national policies to contribute to wider agreement, although the Mediterranean agreements at a regional level and CITES at a global level do point to a growing awareness among some developing countries of the benefits to them of international environmental action, and a willingness to contribute to it. Developing countries have also shown themselves increasingly willing to participate in environmentally related activities of direct benefit to them, such as population planning policies, even when the international rhetoric is unhelpful. The period saw the emergence, and growing incorporation into environmental treaties, of general international environmental principles such as the precautionary and polluter pays principles. It also saw some major failures whose basic common feature, as in the case of the NIEO, UNCLOS and UN Conference on Desertification, was the false assumption that the developed world would be willing to make substantial material transfers to developing countries to help tackle problems by which (in spite of Brandt, the Club of Rome and others) it did not feel directly threatened.

Chapter 6

The Road to Rio

> *Great historical transformations are always bought dearly, often after one has already thought that one got them at a bargain price.*
> JACOB BURCKHARDT

6.1 The second boom

The late 1980s saw the second great boom in popular environmental concern in the West. We have already seen that in the US the environmental policies of the Reagan administration brought about a steady rise in attention to the subject from its low point in 1980. But in 1988-89 there was a sudden leap. For example, the proportion of US opinion poll respondents replying affirmatively to the proposition 'that requirements and standards cannot be too high and continuing environmental improvements must be made regardless of cost' rose from 65% to 80% between mid-1988 and mid-1989.[1] Total membership of the dozen or so major US environmental organizations rose from four million in 1981 to about seven million in 1988 and then leapt to 11 million – with a combined total revenue of more than $300 million a year – by the beginning of 1990.[2] In that year too the majority of respondents to a Louis Harris poll said that a clean environment was more important than a satisfactory sex life.[3]

In the UK the percentage of opinion poll respondents listing the environment as one of the most important issues facing the country, having remained

---

[1] M. Tolba et al., *The World Environment 1972-1992*, Chapman and Hall, London 1992.
[2] B. Bramble and G. Porter, 'NGOs and the Making of US Policy', in A. Hurrell and B. Kingsbury, eds, *The International Politics of the Environment*, Oxford University Press, Oxford, 1992.
[3] *Atlantic Monthly*, October 1990.

at under 10% throughout the 1980s, rose in early 1989 to almost 35%.[4] UK membership of both Greenpeace and Friends of the Earth increased by a factor of six between 1985 and 1989, while membership of the World Wildlife Fund more than doubled.[5] The British Green Party, having polled negligibly throughout the 1980s, suddenly in 1989 got 14.5% of the votes cast in elections for the European Parliament.

A similar bounce in environmental concern at the end of the decade was apparent throughout the developed world. In countries such as France and Italy, where popular concern about the environment had fallen back from its peak of the late 1960s, there was now a strong revival (the French greens polled 10.6% of the votes in the 1989 European Parliament elections compared with 6.7% five years earlier, and even in countries where popular environmentalism had always remained strong (such as Germany, the Netherlands and Scandinavia) the end of the decade saw some reinforcement. In this period green parties polled at record levels in Australia, Belgium, Finland, Germany, Iceland, Italy, Luxembourg, Sweden and Switzerland.[6] In 1988 the Swedish greens became the first new party to enter the Swedish parliament for ten years and Danish membership of environmental NGOs became (presumably due to dual memberships) higher than Denmark's total population.[7] In 1989 environmental dissatisfactions played a major role in bringing down the governments in the Netherlands and Norway.[8]

The one developed country exception to this pattern was Japan where, after the explosion of popular activism in the 1960s and the swift government legislative response, public opinion on the environment, as on so many other issues, had reverted to its normal quiescent state. Japan came bottom in a 1989 UNEP survey of popular environmental concern in 14 countries, and the combined total membership of all Japanese environmental organizations was about the same as that of Friends of the Earth alone in the UK (250,000 members).[9]

---

[4] M. Levy, 'The Greening of the United Kingdom: An Assessment of Competing Explanations', paper delivered at the August 1991 meeting of the American Political Science Association, Washington DC.
[5] Tolba et al., *The World Environment*.
[6] *Keesings Contemporary Archives*.
[7] *The Annual Register: A Chronicle of World Events, 1989*, Longman, Harlow, 1990.
[8] L. Caldwell, *Between Two Worlds*, Cambridge University Press, Cambridge 1994.
[9] H. Maull, 'Japan's Global Environmental Policies', in Hurrell and Kingsbury, *International Politics of the Environment*.

A key resemblance between this second explosion of Western environmental concern and the first in the 1960s is that both took place at times of strong economic growth and prosperity in the industrialized countries. Another source of this second environmental boom can be traced to the mid-1980s, when an increasingly global press brought instant pictures into Western homes of a string of environmental or environmentally related disasters. Thus, in addition to the Bhopal accident in India, 1984 saw drought in Ethiopia, widespread floods in Bangladesh and massive industrial explosions in Cubatao (Brazil) and Mexico City. This was followed in 1985 by famine in sub-Saharan Africa, a cyclone in Bangladesh, an earthquake in Mexico, and the discovery of a vast 'hole' in the atmospheric ozone layer. The next year saw the Chernobyl nuclear accident and a major fire in a Swiss chemical factory which extensively polluted the Rhine (and was promptly dubbed 'ChernoBasle'). In 1987 there was renewed famine in Ethiopia and the worst hurricane for a century at least in the UK.

It was in 1988 that the world took on an almost apocalyptic hue (at least as seen in the pages of the Western press). This year joined 1983, 1987 and 1981 as four of the five hottest years since records began, thus contributing to widespread concern about the possibility of global climate change. These concerns were reinforced by a number of climatic disasters: one of the worst ever droughts in the US midwest (bringing huge agricultural losses and a 20% fall in the US wild duck population), one of the most powerful hurricanes of the century in the Caribbean and Mexico, and extensive flooding in Bangladesh. In addition, Western countries found themselves charged with dumping toxic waste in Africa, and one toxic waste ship – the *Karin B* – was refused entry in six European countries before having to return to its port of origin in Nigeria. Two separate fatal epidemics killed thousands of Atlantic dolphins and thousands of Baltic seals. Algal blooms wiped out extensive tracts of marine life off the coasts of Norway, Sweden and Denmark. At the UN, Soviet foreign minister Edward Shevardnadze called for an 'international regime of environmental security'. At the end of the year a leading Brazilian rainforest campaigner, Chico Mendes, was assassinated. *National Geographic* magazine devoted the whole of its 100th anniversary issue to the problems of the environment; *Time* magazine repeated its innovation of 1970 by replacing its 'Man of the Year' by 'Planet of the Year – the Endangered Earth'. While it is not clear whether this was a genuinely unusual string of disasters or whether it merely looked like it as dramatic coverage of each event fed even more dramatic coverage of the next (on the 'issue/attention' model noted in Chapter 2), the effect was to generate a sharp rise in press and popular environmental concern.

This time, the problems which principally attracted the attention of environmentalists and the press were international problems – the ozone layer, climate change, third world drought and deforestation – even though there is good evidence that Western public opinion was still most interested in domestic issues. For example, in a 1989 poll in the UK respondents ranked acid rain and global warming only in the middle of a list of 21 issues.[10]

Apart from the sheer quantity and global reach of the press coverage, two other factors at this time gave impetus to international governmental discussion of the environment. The first was a growing perception in the third world that this latest burst of Western concern about the environment might be translatable into Western economic assistance to developing countries (in a revival of the 'North-South environmental bargain' ideas last seen at the NIEO). At the 1987 UN General Assembly, the prime ministers of India and Zimbabwe argued that developing countries often had to exploit the environment in order to survive, and that the solution to this problem was an extra injection of aid to help remedy conditions which made exploitation of the environment inevitable.[11] The second was the publication in the same year of yet another 'wise men's' report – but one which has exercised strikingly more influence than its predecessors.

This report was *Our Common Future*, prepared by the World Commission on Environment and Development chaired by the Norwegian prime minister, Mrs Brundtland.[12] This commission had been set up in 1983 following an initiative at the 1982 'tenth anniversary' session of UNEP. Much of the content of the report follows firmly in the footsteps of previous such documents. There is the traditional emphasis on global interdependence, equity, more help for developing countries and natural resource limitations on economic growth. The report does, however, show a hard-headedness uncharacteristic

---

[10] Levy, *The Greening of the United Kingdom*.
[11] *New York Times*, 20 October 1987.
[12] World Commission on Environment and Development, *Our Common Future*, Oxford University Press, Oxford 1987.

of such exercises in the emphasis it gives to institutional factors; in particular arguing that decision-makers (for example in finance ministries or agriculture ministries) should be held accountable for the environmental consequences of their decisions. But the genius of the piece lies in its adoption and promulgation of the concept of 'sustainable development', that is, development which 'meets the needs of the present without compromising the ability of future generations to meet their own needs'. The concept of sustainable development thus effectively bridged the intellectual and political gap which had been apparent at least since Stockholm between those (particularly in the developing world) arguing for economic growth, and those (particularly in the developed world) arguing for environmental protection. It encouraged growth, but incorporated environmental concern in order to ensure that growth should not ultimately undo itself. In one neat formula, Mrs Brundtland had provided a slogan behind which first world politicians with green electorates to appease, and third world politicians with economic deprivation to tackle, could unite. The formula was of course vague, but the details could be left for later.

So a combination of circumstances – climatic and other disasters, growing scientific knowledge, the Brundtland Report – brought the international environment to the top of the Western political agenda by the late 1980s. The political potency of environmental issues, even at this time, should not however be exaggerated. Although they played a prominent role in the campaigns for both the UK general election of 1987 and the US presidential election of 1988, and although both Mrs Thatcher and Mr Bush (as President Reagan's

Vice President) had difficult environmental records to defend, and were opposed by most environmental groups, both won. Nevertheless, as in the late 1960s, Western politicians were obviously under significant pressure to be seen to be responding to the array of international environmental issues confronting them.

6.2 Chernobyl

Quite a lot of this response was piecemeal and reflects the pattern we have already seen of disasters driving forward the manufacture of international environmental law. Perhaps the most dramatic example is provided by Chernobyl, the town in Ukraine where, on 26 April 1986, a sequence of operator misjudgements caused the worst civil nuclear accident in history.[13] Twenty people were instantly killed, 116,000 had to be evacuated, and a cloud of radioactivity spread across 21 nations of West and Central Europe. The total eventual count of premature deaths from radiation sickness and cancer is estimated to run into thousands. Even now, eight years later, some sheep in Wales, over 1,000 miles from the site of the accident, are too radioactive to be placed on the market.

The Soviet response to the accident was to conceal it until measurements in the West made it undeniable. We have noted above the difficulty of achieving international agreement on nuclear safety issues. In this case, however, as in that of the *Torrey Canyon*, the political need was imperative and the gaps in the international system glaring. After a brief skirmish within Western Europe about where work should be pursued (with the countries more hostile to nuclear power preferring the EC, where they had a large voice, while the major nuclear powers preferred the IAEA, where they had the dominant influence) a swift negotiation in the IAEA produced conventions on assistance in the case of a nuclear accident and early notification of a nuclear accident, both of them negotiated within six months of the explosion, and both of them now signed by over 70 countries and ratified by over 40.

The Chernobyl accident also forced the European Community (after an epic argument between pro- and anti-nuclear states which lasted until 1988) to establish agreed radioactivity standards for foodstuffs, which have since

---

[13] *The Annual Register: A Chronicle of World Events, 1986*, Longman, Harlow 1987.

been widely accepted internationally. The interesting feature of these standards is that those finally agreed were much tougher than those originally recommended by the Community's scientists, because it was politically impossible for some member states to appear 'lax' on the issue.

Undoubtedly the main, and probably most enduring, product of the accident is that it significantly boosted Western public opposition to the construction of new nuclear plants (in the case of West Germany from 44% to 82% of the population). It enabled the antinuclear movement to claim that its fears had been vindicated,[14] and in most of the developed world it has made still more difficult the prospects for an industry already suffering from very high costs and widespread public mistrust.

6.3 Toxic waste

The international reaction to evidence that Western toxic waste was being dumped in Africa was similarly vigorous. In the course of 1988 Nigeria found 8,000 drums of toxic waste at its port of Koko (photographs of which, with children playing among them, duly appeared in the world's press) and charged Italy with having dumped them there; and Guinea arrested the representative of a US firm which deposited fly ash on an island near the capital.[15] These two countries thus put on display the seamy side of the estimated $3 billion per annum trade in toxic waste from developed to developing countries.

Plainly, this was an issue with all sorts of imperialistic overtones, and a meeting of the Organization for African Unity in May 1988 in Addis Ababa had little trouble in agreeing to condemn the practice. That view was shared by many in the West, such as the Dutch minister of the environment who referred publicly to 'waste colonialism'.[16] The general Western state of environmental alarm that summer was reinforced by the *Karin B* incident, mentioned above, where a string of West European governments found themselves reassuring an agitated public opinion that they would find means to prevent

---

[14] Christopher Flavin, *Reassessing Nuclear Power: The Fallout from Chernobyl*, Worldwatch, Washington DC 1987.
[15] M.A. Montgomery, 'Travelling Toxic Trash: An Analysis of the 1989 Basle Convention', *The Fletcher Forum*, summer 1990.
[16] G. Porter and J. Brown, *Global Environment Politics*, Westview, Boulder Co 1991.

the ship docking and unloading on their shores. This upsurge of interest in the issue fell in well with the agenda of UNEP, which had in any case already been studying the world toxic waste trade with a view to regulating it.

The upshot was a UNEP-hosted negotiation which in the five months from November 1988 to March 1989 produced the Basle Convention on the Transboundary Movements of Hazardous Waste. The positions taken in this negotiation were revealing in a number of ways. First, this was probably the first major international environmental negotiation in which the developing countries, led by the Africans, were demanding tougher environmental regulation than the West. The wave of developing country anxiety following the Nigeria and Guinea cases (Nigeria, of course, being particularly influential in the OAU), coupled with the exploitative overtones of the subject, outweighed any commercial interest in the reprocessing of waste and produced an African demand that the trade be internationally banned.[17]

The West, on the other hand, conscious of the growing volume of hazardous waste it was producing, and having just been sharply reminded of the difficulty of finding disposal sites at home (another simultaneous example of which was the fact that one of the first consequences of the liberalization of Poland's economy was a string of contracts for the importation of toxic waste, described as 'raw materials', from the West[18]), offered no more than a regime of 'informed consent' under which importing countries had to be properly notified before shipments could take place. In the absence of any Western flexibility, this was the system finally adopted.

It is noteworthy that over the same period the European Community was moving towards a much more restrictive philosophy with regard to waste movements between its own member states (in what is notionally a free market) and has also agreed in the fourth of the Lomé agreements (see Chapter 5) to ban waste exports to the group of developing countries with which it is linked through the Lomé process. Indeed, the prominence of environmental issues in general, and waste issues in particular, at this time was such that the

---

[17] This demand of course emphasized the lack of confidence of developing countries in their own environmental regulatory systems through which they could otherwise simply have stopped the import of such waste (and indeed, a group of them have since negotiated an agreement, the Bamako convention to ban such imports).

[18] *World Resources 1992-93*.

argument about waste exports nearly sank the whole Lomé IV negotiation (which in principle was much more about aid than about the environment), with the Community countries only agreeing to ban waste exports to the developing countries when the latter agreed to ban waste imports from *all* developed countries.

Altogether, Basle gives the impression of having been characterized even more than most such negotiations by a wide divergence between the ostensible environmentalism of the developed country participants and their actual political and economic motivations. The developing countries did not get the ban on toxic waste exports they had demanded, but even in conceding an 'informed consent' regime the West was making it more difficult for itself to dispose of its toxic waste overseas . The convention received its 20th ratification, and thus entered into force in 1992, but with a great deal less enthusiasm and (in particular African) support than seemed likely when the negotiation had been launched four years earlier. The achievement of a level three response to the toxic waste problem is likely to depend not so much on the Basle convention as upon whether developing countries effectively enforce their own domestic legislation on the subject and implement the Bamako convention.

6.4 Antarctica

Another issue visibly blown off course by the gale of Western environmentalism in late 1988 was the (thitherto highly obscure) management of Antarctica. Antarctica is environmentally unique, not only in terms of its physical conditions and the species that live there but also in terms of its remoteness from the activity, and pollution, elsewhere on the planet. This meant, for example, that it was in Antarctica that the ice cores were drawn which have given us the best picture of the earth's atmosphere in earlier ages, and over Antarctica that the full extent of the depletion of the ozone layer first became apparent. Due to conflicting territorial claims Antarctica is administered through regular meetings among the 38 'Antarctic Treaty states'. Over time, they have erected an effective network of treaties to protect the Antarctic environment, but have also kept open the possibility of exploiting the potentially vast mineral and oil reserves of the continent. In June 1988 the Antarctic Treaty states crowned a decade of negotiation with agreement on the Convention on the

Regulation of Antarctic Mineral Resource Activities (CRAMRA), one of whose central aims was precisely to impose stringent environmental controls on any mineral prospecting and extraction in Antarctica.

The convention was promptly denounced, however, by Western environmental NGOs as a 'miner's charter'. In May 1990 Australia and France (both of them facing significant domestic green political pressures[19]) formally rejected CRAMRA and proposed that Antarctica should instead be made a 'world park' in which mineral exploration would be banned.

One by one the other Antarctic Treaty states – offered a choice between the real political costs of offending aroused domestic environmental opinion and the highly theoretical economic benefits of eventual exploitation of Antarctic minerals – fell more or less grudgingly into line with the Franco-Australian proposal. Thus at a meeting in Madrid, just three years after CRAMRA had been unanimously approved, it, and the ten years' work that had gone into it, were junked and replaced by a 50 year moratorium on all mineral-related activity in Antarctica. This turnabout was a quite remarkable demonstration, in the heightened environmentalist atmosphere of the late 1980s, of the power of public environmental concern to change official positions in a wide spread of states when there were not pressing economic reasons for acting otherwise.[20]

6.5 The ozone layer

Background: before the 'hole'

While the issues raised by Chernobyl, the toxic waste trade and Antarctica all reflected in some way the new-found political potency of the environment, undoubtedly the biggest, most difficult and most influential negotiation of the period was that on the ozone layer.[21] Good accounts of the science of this issue can be found elsewhere.[22] Briefly, scientists began to suspect in the mid-

---

[19] French, *After the Earth Summit*.
[20] Porter and Brown, *Global Environment Politics*.
[21] The fullest account of this negotiation is Benedick, *Ozone Diplomacy*, but this is marred by its one-sidedness. For redress see F. McConnell's review of the book in *International Environmental Affairs*, Summer 1991. See also E. Parson, 'Protecting the Ozone Layer: The Impact of International Institutions', in P. Haas et al., eds., *Institutions for the Earth*, MIT Press, Cambridge MA, 1993.
[22] See e.g. Tolba et al. *The World Environment*.

1970s that a number of artificial and extensively used chemicals were eating into the layer of ozone (a special form of oxygen) which surrounds the earth's atmosphere and which prevents a great deal of ultraviolet radiation from the sun from reaching the earth's surface. If this ultraviolet got through it would cause increased animal and plant cell mutation and in particular cancer in humans. The best known of the offending chemicals are chlorofluorocarbons (CFCs) which are widely found in fridges, air conditioners and aerosols.

These suspicions came hard on the heels of an earlier round of concern about the ozone layer which, some scientists had hypothesized, would be damaged by nitrogen oxide emissions from supersonic aircraft. Subsequent work has shown that those earlier scientific fears were excessive, but they were extensively used in debates in the US to argue first that a supersonic aircraft should not be manufactured there, and second that the Anglo-French supersonic aircraft Concorde should not be granted US landing rights. We have already seen that there were attempts to raise this issue at Stockholm. This use of what later proved to be exaggerated scientific concern about ozone depletion in an attempt to influence an important government decision in another area (as many of the financial prospects of Concorde depended upon its being able to land in the US) caused real difficulty later. When the ozone issue came back with CFCs replacing supersonic aircraft as the villain of the piece there were strong suspicions, notably in France, that once again this was a US search for economic advantage dressed up as environmental concern.

Nevertheless, the new ozone depletion theory was, like the earlier 'limits to growth' predictions of the Club of Rome, sufficiently apocalyptic to attract a significant measure of press and public interest in North America and Western Europe. It provoked a great deal of scientific work and debate, the results of which at that time were far from conclusive. Scientific estimates made in the period 1974-83 of the proportion of the ozone layer likely to be depleted oscillated wildly in their results from 3% to 19%. The big chemical manufacturers, notably Dupont Corporation, unsurprisingly argued that the science of the CFC-ozone depletion link was too uncertain to justify action against a $3-5 billion dollar a year industry, with the investment and employment consequences that would entail. But all the big chemical companies at the same time launched 'precautionary' research programmes on possible CFC substitutes. On the other side, the environmental groups took up the CFC theory and argued, in particular, that the use of CFCs in aerosols (about 50% of the US and world markets) was 'inessential and frivolous' given the threat that it entailed, and should be banned. It is a testimony to the political potency in the USA of even unsubstantiated environmental fears that despite the continuing scientific uncertainty, and despite the very large economic stakes involved, such arguments prevailed and led in 1979 to a US ban on CFCs in aerosols, followed by similar bans in Canada, Norway and Sweden.[23]

The outcome in Western Europe was different. Even there, the level of popular and environmentalist concern, combined with US demands for parallel action, required some appearance of a response. But consciousness of the uncertainties and costs produced agreement in 1980 on an EC 'cap' on CFC production (as well as an informal agreement to cut aerosol use by 30%) which in practice would have allowed production to continue to rise to the end of the century. In Japan, too, there was no significant popular concern and it was only in response to sharp international criticism that the government announced, in 1980, that it would 'work to freeze [CFC] production capacity'.[24] The developing world, negligible producers and consumers of CFCs, took no visible part in the debate or interest in the issue.

The combined effect of these measures was to reduce world CFC production by about 25%, take the public urgency (where it had existed) out of the issue, and shift such continuing action as there was to international fora. Indeed, further US domestic regulation was blocked in 1981 through a vigorous public campaign by an industrial lobbying group, one of whose central points was that further action should be at the international, not unilateral, level. In the absence of any potential demand for CFC substitutes the research programmes launched by the chemical companies (which had in fact

---

[23] It can of course be justified on the basis of the 'precautionary principle' that one should not wait for full scientific proof before acting. The question then becomes how much scientific evidence is enough, and here it is noteworthy that the vast majority of OECD governments, including some of the scientifically best advised (France, the UK, Germany) concluded that a ban was not justified.

[24] M. Schreurs, 'International Cooperation for the Environment in a State Centred World: Comparing the Responses of Japan and Germany to Global Environmental Threats', Centre for Science and International Affairs, Kennedy School of Government, Harvard, 1993, preprint.

shown that substitutes could, at a price, be produced) were not pursued beyond the test tube stage, and their results were filed away.

The Vienna Convention

UNEP had in fact taken an early interest in what seemed an international environmental issue *par excellence* (even though a number of countries had responded to a 1981 UNEP circular saying that they did not use CFCs so 'their ozone layer' was not threatened[25]). It followed its traditional approach of initially bringing together international groups of scientists to move towards a shared understanding of the risks. The late 1970s saw a sequence of efforts by those (notably the Scandinavians) who had made serious cuts in their CFC use to start an international negotiation whose evident aim was to get others to make similar cuts. Finally, in 1981, a negotiation was launched under UNEP auspices, but with those who had not introduced stringent controls viewing the aim as being no more than to formalize cooperation on research and data exchange.

> **The most important effect of the Vienna Convention on Protection of the Ozone Layer (as with LRTAP) was that it provided for continuing discussion of the issue .. a number of factors combined to bring about a dramatic change of stance in Europe and Japan .. the resulting Montreal Protocol on Substances that Deplete the Ozone Layer is epoch-making in a number of ways.**

This period of the negotiation saw no hardening of the science, nor any renewal of public concern; indeed, an erroneous 'chlorine catastrophe' theory of 1984-5, which predicted colossal ozone losses but was swiftly shown to be mistaken, added significantly to the scepticism with which the scientific basis of the whole issue was viewed, at least by those countries which had no vested interest in pressing others to take tougher measures. So, roughly speaking, the negotiation produced the results the minimalists were seeking. After

---

[25] Parson, 'Protecting the Ozone Layer', mimeo, Harvard University.

three years of unproductive wrangling in which each major country or bloc argued, essentially, that the international community should adopt whatever measures that country or bloc already had (or had not) introduced, the final convention was signed in Vienna in March 1985 by 20 nations (virtually all of them developed, and not including Japan). It provided only for research and monitoring cooperation, as well as information exchange on CFC production and emissions. The disagreement on control measures was, in the traditional way, resolved by remitting the subject to subsequent discussions. Indeed, it is arguable that the most important effect of the Vienna Convention (as with LRTAP) was that it provided for continuing discussion of the issue.

A change of momentum

The period between the signing of the Vienna Convention and the resumption, in December 1986, of negotiations on control measures saw a number of crucial developments. Scientific opinion, through a major assessment exercise conducted by UNEP and the World Meteorological Organization and a subsequent international scientific workshop, moved closer to a consensus that CFCs did constitute a significant threat to the ozone layer. Undoubtedly the most dramatic scientific development was the discovery in 1985, by a British team of scientists, of the Antarctic ozone 'hole' – a vast region (larger than the continental US) over Antarctica where ozone levels fell to about half their historic values for three months a year. This level of ozone depletion was so much greater than anything that had been predicted that the British team had initially doubted their results, and US and Japanese teams had rejected similar readings as obvious errors.

The hole attracted widespread press and public attention in Europe and the US. It also had the (political) virtue of being recurrent, thus reviving and accentuating concern each year as it got bigger. Moreover, it constituted a very concrete and vivid public image of events in the stratosphere in what had hitherto been a rather arcane scientific debate. There seems little doubt, therefore, that even before scientific confirmation of its direct relevance to the CFC debate (which was in fact not definitive until after the negotiation was over) the hole significantly altered the political background to the negotiation in much the same way as Chernobyl energized international action on nuclear safety issues.[26]

Other developments prior to the resumption of negotiations took place on the political and industrial levels. Dupont, having lost the 1977 battle about the US ban on CFCs in aerosols, saw which way the wind was blowing and in 1986 resumed research into substitutes for CFCs. This fact, and Dupont's view that substitutes could be developed, of course became known and were an additional arrow in the CFC abolitionists' quiver. Indeed, one argument aired in developing the US position for the negotiations to come was that only a sharp limit on CFC production would create the incentive for companies to develop alternatives, and the market for these alternatives once available.

The US position was also affected by a hardening of scientific views on the CFC-ozone depletion link. Subsequent events suggest that those elements in the US government which might have been expected to argue against a radical US position on CFC cuts did not do so, either through inadvertence or because they expected the negotiations to make little progress given the lack of enthusiasm for severe cuts in other countries, notably Western Europe. The result was that the US formal negotiating position emerged as a demand for a freeze followed by timed moves to a more or less total ban on CFC production. Given the weight and political influence of the US, and the energy with which this position was pursued, there is no doubt that it had a major effect on the final outcome.

In Europe the situation was different. Popular concern, although rising, was less intense than in the US (except in Germany and Scandinavia). The science – even with the ozone hole – remained unproven. European industry had not faced even an aerosol ban and had therefore done no work on substitutes beyond the 'test tube' research of the 1970s. Finally, the European countries worked through the EC on this issue; in the Community, as in many multilateral organizations, a position once fixed is often difficult to change, and the current EC position was opposition to controls. Indeed, there was a widespread suspicion in the EC that the main US motivation in pursuing the issue was to gain industrial advantage. US industry, seen as having adapted to a domestic ban on CFCs in aerosols, and known to be working on substi-

---

[26] Although Benedick, *Ozone Diplomacy*, mysteriously denies this.

tutes, was suspected of using public environmental concern to try to steal a march on its competitors. In Japan the domestic pressures for action were even weaker than in Europe.

The Montreal Protocol and London Amendments

The progress of the negotiation of what was to become the Montreal Protocol was therefore quite remarkable. After a predictably slow start in which the EC in particular stuck firmly to its position of agreeing to no more than a freeze in current CFC production capacity, the Community (and with it Japan) were edged over the six months from February to September 1987 to agree to a timetable for the 50% cut in CFC production which was the final outcome. This also, of course, required a substantial shift on the US side from their original call for a near-total phase-out, a shift that was no doubt made easier by a last-minute attack on the US official negotiating position by certain US agencies which had not expected the negotiation to make as much progress as it had.

A number of factors combined to bring about this dramatic change of stance in Europe and Japan. There was, of course, intense pressure from the US. This was exercised through both political and public channels; in particular, US negotiators publicly criticized European immobility and US NGOs encouraged their European counterparts to campaign on the issue.[27] In Japan there was concern that if they did not shift they might be subject to US trade sanctions.[28] Second, there was disunity within the Community itself, with Germany in particular (with an electorate sensitized to international environmental issues as a result of the acid rain debate, and strong Green Party representation in parliament) demanding that the Community go further, and threatening, if necessary, to break Community ranks if it did not. Perhaps the crucial factor which helped to change European ministers' minds was that the science firmed up significantly while the negotiation was under way. This also affected European industrial attitudes, since industry scientists were part

---

[27] Though with limited effect. A UK government minister at one point chided the environmental NGOs for their inattention to the subject. See McConnell, review of *Ozone Diplomacy*, op. cit. footnote 21.

[28] Schreurs, 'International Cooperation for the Environment'.

of the scientific community analysing the data.²⁹ In particular, an international scientific meeting at Wurzburg gave quantitative values to the amount of damage they felt the various chemicals under debate were doing to the ozone layer. The prospects of agreement were also enhanced immeasurably by the technique of the UNEP executive director, Mustafa Tolba, of getting just the key participants together in a room and hammering out with them a single text on the basis of which to push the negotiation forward.

The resulting agreement, the Montreal Protocol on Substances that Deplete the Ozone Layer, signed by 24 countries in Montreal in September 1987, was epoch-making in a number of ways. It was the first treaty ever in which countries agreed to impose significant cost on their economies in order to protect the global atmosphere. It was achieved despite continuing uncertainty on both the science of and the costs imposed by ozone depletion.³⁰ Despite being a hard-fought, and in scientific terms untenable, political compromise (the 50% cut was either far too much, if CFCs were not causing the Antarctic ozone hole, or far too little, if they were),³¹ it marked a political and psychological breakthrough, at least among developed countries, with regard to CFC regulation. Following scientific confirmation in early 1988 that CFCs *were* responsible for the ozone hole, the major chemical companies (again led by Dupont) announced their intention to phase out CFC production totally, and their governments agreed at the follow-up meeting in London in 1990 to a complete phase-out by the end of the century.

The Montreal Protocol raised two big new sets of international questions. The first was its imposition of restrictions on trade in CFCs and CFC-related products between parties and non-parties to the protocol (thus following CITES). The aim of this provision was to encourage countries to sign the

---

[29] ICI scientists in particular had access to the work the British Antarctic Survey was doing on the Antarctic hole at precisely the time of the Montreal negotiation, and were able to let company and government decision-makers know the likely results.

[30] Although estimates were appearing at about this time which suggested that the costs of ozone depletion could be enormous. A later UNEP publication (*Economic Panel Report*, 1989) estimated total costs in the US, including skin cancer deaths valued at normal compensation rates, at $3,517 billion by AD 2075.

[31] Untenable except, perhaps, on the basis of the 'precautionary principle' – but any level of cut, from 1% to 99%, could have been justified on that basis. The point about 50% is that its only real justification was political.

protocol; but, despite the testimony of GATT lawyers in Montreal, it looks difficult to square with global free trade arrangements and raises the whole issue of how those arrangements should be adapted to take the environment into account.

The second prophetic feature of the protocol was the role played in it by the developing countries. Developing country concerns had been of marginal significance in the Montreal negotiation. Few developing countries were present at the early sessions and none was included in Tolba's 'key group'. The provision written into the protocol to gain their adherence was the right to a ten-year delay in phasing out CFCs by developing countries as well as highly unspecific references to their need for financial and technical assistance. A number of relatively industrialized developing countries – notably Mexico, Venezuela, Egypt, Kenya and Thailand – were sufficiently persuaded by these concessions, and by the trade measures, to sign the protocol and to start to limit their CFC consumption.

However, it rapidly became clear following Montreal that this was not enough. Although developing country consumption of CFCs was small (less than one-twentieth per capita of developed country consumption), it was growing rapidly – particularly in China and India, both of them non-signatories and both of them huge potential producers and consumers of CFCs. It was quite clear that if the threat to the ozone layer was really to be tackled then the developing countries must form part of the effort. By the end of 1989 only 21 developing countries had signed the Montreal Protocol and a number of others, notably India, were making it clear that they would only get involved if the West would provide the extra finance and technology necessary for them to develop industry based on the (much more complex and expensive) CFC substitutes rather than CFCs. These demands of course caused problems in the West, where overseas aid is always under financial pressure and where technology is usually in the possession of private companies which usually have their own priorities and prejudices about the countries to which it can and cannot advantageously be supplied.

It was these North-South tensions which dominated the run-up to the 1990 London meeting scheduled to review the Montreal Protocol. The US. concerned that any funding arrangement for the developing countries in the ozone layer context might become a precedent for the (much more expensive) issue of climate change, nearly wrecked the meeting by announcing two months before it that any finance should come from existing aid resources, but was then persuaded to pull back from that position. London itself saw an extremely tense contest between, on the one hand, the determination of the major Southern states to get the best possible financial and technological terms for their participation and, on the other, Northern political determination to get the South on board without taking on vast and open-ended financial and technological commitments. The upshot was agreement to establish a new fund, initially of $160-240 million (the difference being a $40 million tranche each for India and China if they acceded, as they eventually did), to be contributed by developed countries to help developing countries cease to use CFCs. A key issue proved to be decision-making powers over the raising and spending of the money, which were very carefully balanced between North and South.

The agreement on technology transfer was even more difficult to achieve and required a number of informal conversations between Southern environment ministers and Northern chemical companies to reassure the former that CFC substitute technology was likely to be forthcoming in the right circumstances. The text in the agreement notes carefully that developing countries' 'capacity to fulfil' their obligations 'will depend upon effective implementation of financial cooperation... and transfer of technology', thus leaving judiciously unclear the extent to which developing countries are bound by the agreement if they feel that adequate quantities of finance and technology have not been forthcoming.

Although international work on protecting the ozone layer has carried on beyond the London meeting – deadlines have been tightened, more chemicals are being phased out – all of the essential features were in place at the end of that meeting. Since then, all the major developing countries have joined the process and the Montreal Protocol now has 74 ratifications.

Conclusions

The process and outcome of the ozone negotiations exemplify many of the key features of modern international environmental business, and were taken very much as a model in the climate change negotiation, to which we turn in the next chapter. Several points are particularly noteworthy, beginning with the role of science. Disagreements about the science undoubtedly held back international action; but the mere existence of scientific disagreement was not decisive. The disagreements narrowed in the course of the ozone layer negotiation but were still real when political and popular pressures nevertheless brought about agreement on the Montreal Protocol. One factor which contributed, perhaps decisively, to the build-up of those pressures was the discovery of the ozone hole – a popularly accessible, concrete image of the environmental damage which needed to be cured. And the whole history of the negotiation has given considerable extra persuasiveness and political salience to the 'precautionary principle'. It has subsequently become very clear that Governments by acting ahead of absolute scientific certainty have avoided costs and environmental damage which they would otherwise have had to bear, and that they could have avoided still more by still earlier action.

In second place, the role of industry was significant. There was a constant interplay and contact between governments and their chemical industries as the negotiation proceeded. In the early stages of discussion (in effect up to Montreal) the chemical companies carried out 'precautionary' research into substitutes but on the whole lobbied against CFC phase-out. Following Montreal – partly in response to the new regulatory environment which was clearly coming, partly in response to the new clarity of the science (with which they were closely in touch, and which raised the possibility of serious product liability suits in the US if they continued to produce CFCs), and partly with an eye to the higher profits derivable from CFC substitutes – they became firm advocates of total phase-out. Governments were aware of these views and of course took them seriously in formulating national negotiating positions. Even the US held out for just an aerosol ban (to which its industry had already adapted) in the run-up to Vienna, and only went for a total CFC phase-out when Dupont had said it would be able to produce substitutes. However, industrial views were not decisive. The US got well ahead of its chemical industry in formulating its demand for a 100% phase-out of CFCs in 1986, and Europe and Japan ahead of theirs in the Montreal Protocol negotiations of 1987. Again, in both cases the pressures for movement were political and popular. Moreover, it is striking that it was virtually only the chemical industry (which is likely eventually to benefit from the outcome of the negotiation as consumption moves from CFCs to higher-priced substitutes) which made sure that its views were known to government and its interests taken into account. The annual value to the chemical industry of CFC production is only $3-5 billion. Far larger is the 'user' industry (refrigerator manufacturers, electronics companies, etc.) which is responsible for the $300-400 billion worth of capital equipment in the world dependent upon CFCs. It is interesting, therefore, that this sector played virtually no part in the negotiation, and is only now waking up to its implications, and the vast potential costs for them.[32]

Third, there is the North-South dimension. In this case, as often, initial concern and action came from the North, with its large and better-equipped scientific community and higher level of popular environmental sensitivity. Southern involvement, at least for a number of key countries, was made contingent on the supply of Northern finance and technological help. Northern willingness to provide these stemmed from a consciousness that Southern involvement was vital, coupled with the fact that the sums and arrangements needed were modest by comparison with overall international aid flows.

Fourth, there is the institutional dimension. This has two aspects. First, UNEP played a key role in generating international activity on the subject, putting scientists and others together and, as we have seen, getting results out of the negotiations themselves. Second, the process has produced a whole new international institutional structure with regular meetings of experts, officials and ministers, a permanent secretariat, arrangements for information and data exchange, and an international fund.

Fifth, and finally, the exercise must be judged at this stage as a major international environmental success. The vast majority of the world's countries are on board, including all the key ones. The major ozone-depleting chemicals are well on the way to extinction, and as others are being identified they are added to the list. The quantity of CFCs and other chemicals already in existence means that it will be decades before the ozone layer really does begin to recover but, provided the treaty does not collapse over, for example, renewed disagreements about finance and technology transfer, the human race has demonstrated that it can act together to produce a complete level three response to a global environmental problem.

---

[32] See e.g. *Economist*, 29 January 1994.

6.6 The Role of Business and Industry

We have noted the important impact on the ozone layer negotiations of the views and lobbying of industry. This, of course, is not a phenomenon unique to stratospheric ozone. From very early in the history of international environmental cooperation, industries likely to be affected by the outcome of any particular negotiation have ensured that governments (and often the public) are aware of their views, and have lobbied to ensure these are taken into account. In general, too, industry has been highly conscious of the likely costs of environmental regulation and has tended to oppose it. Thus at the beginning of international action to deal with marine oil pollution in the 1930s the shipping and oil industries resisted requirements for new pollution control equipment to be fitted to tankers, and continued to do so when the negotiations resumed in the 1950s and 1960s.[33] We have seen how in the 1970s the Italian government resisted the inclusion of titanium dioxide in the Mediterranean marine dumping protocol because the Montedison Company wished to continue to dump. In the European acid rain negotiations of the 1980s the British Central Electricity Generating Board (which was nationally owned) kept the British government very fully informed on the high costs of any agreement which would require serious cuts in British sulphur dioxide emissions. Similarly, in 1988 the Peugeot car company made a last-ditch effort to get the French government to block EC agreement on stringent exhaust emission standards for small cars. In the ozone layer negotiations themselves we have seen that the differing speeds at which the American and European CFC manufacturers came to accept that CFCs had to go correlate well with the differing speeds at which the US and European governments accepted those same facts.

This long and unsurprising history of industrial opposition to environmental regulation, coupled with the association of big industrial names with some major environmental disasters – Union Carbide with Bhopal, Exxon with the *Exxon Valdez*, the nuclear industry with Chernobyl, and so on – has left big industry, or at least major sectors of it, with a villainous reputation among some environmentalists. The antipathy has often been mutual, as many industrialists have seen excessive environmental regulation (brought on by en-

---

[33]R. Mitchell, 'International Oil Pollution of the Oceans' in Haas et al eds., *Institutions for the Earth*, MIT Press, Cambridge MA 1993.

vironmentalist fervour) as a threat to competitiveness. But as environmental concern has increasingly become an uncontentious 'motherhood' value in Western society, and as industry (which is after all part of society) has adapted, so the 'industry vs environment' picture of the relationship has become something of a caricature.

Industrial adaptation to increased popular environmental awareness started relatively early in the history of Western environmentalism. We have already seen that following the *Torrey Canyon* accident in 1967 the major tanker owners established voluntary systems, TOVALOP and CRISTAL, to ensure that those affected by such accidents in the future would be properly compensated. There was of course self-interest in this – the owners did not wish to be confronted with an unmanageable assortment of different liability systems imposed by different coastal states – but it was enlightened self-interest in that their reaction to the problem was not to resist liability systems (which would in any case have further damaged the environmental reputation of the industry) but rather to create one that they could live with.[34]

This pattern of business being prompted by self-interest to pay attention to the environment has been frequently repeated in the two ensuing decades. At the very minimum, business has had to adapt to a growing web of environmental rules and regulations. As environmental consciousness has grown, penalties for infraction have grown stiffer. Fines (originally often regarded as a normal business expense, and still so regarded in many areas of the third world) have grown steeper, and in the US managers are now regularly sent to jail for environmental offences. In 1990, for example, the Environmental Protection Agency (EPA) alone brought cases which resulted in nearly $30 million of fines and a total of 22 years of jail sentences.[35] The threat, too, particularly in the US, of multi-million-dollar liability suits over environmentally damaging products and practices has compelled companies to look very carefully at how exposed they are.

Over and above simply obeying the law and avoiding civil liability, the need for Western business to *look* environmentally concerned has now become

---

[34] M'Gonigle and Zacher, *Pollution, Politics and International Law*.
[35] S. Lamb, *The Greening of Business in the United States*, unpublished report prepared for the Department of the Environment, London 1992.

almost universal. Given an environmentally aware public with a plethora of consumer choice, a company can pay a significant price for being branded a polluter. Recent years have also seen the emergence of and acquisition of some market power by the 'green investor' – the individual or institution which takes environmental performance into account in deciding which companies to support or invest in.[36] Thus Exxon suffered a substantial loss of sales after the *Exxon Valdez* disaster, and one of the pressures on the British chemical industry in the final rounds of the Montreal Protocol negotiation was a consumer boycott of CFC-using aerosols. British Nuclear Fuels Ltd have had to lay on a major campaign of openness at their reprocessing plant at Sellafield in order to offset its earlier accident-prone reputation. Large industries are now spending real money in order to protect their green image. A number of US power companies have bought, or planted, a chunk of forest when they have opened a new power station in order to offset the carbon dioxide emissions. Mitsubishi, which has been particularly targeted in campaigns against Japanese industrial consumption of wood from the rainforests, is now financing a major experiment in rainforest regeneration.[37] McDonald's has altered its packaging to minimize environmental impacts, and the British do-it-yourself company B&Q stipulates that all the timber it sells must have been sustainably forested.

Beyond this widespread 'defensive greenness', more restricted industrial sectors have gone further in trying to turn the environment to commercial advantage, either through cutting the resources (such as energy) they consume or through expanded sales. The most visible activity of this sort is of course catering to the 'green consumer'. This is a fast-growing market in the West. The number of Britons buying a product for environmental reasons rose from 19% in 1988 to 42% in 1989 and the *Green Consumer Guide* was 1988's unlikeliest UK bestseller.[38] Major manufacturers and retailers such as AEG in Germany and Body Shop in the UK have made the greenness of their products a key selling point. Whole product areas have been affected. The proportion of phosphate-free detergents in the world market has risen from 6% in the 1970s to 60% in 1988. Nor, of course, is the 'green' market only a

---

[36] F. Cairncross, *Costing the Earth*, Economist Books, London 1991.
[37] S. Schmidheiny, *Changing Course*, MIT Press, Cambridge, MA 1992.
[38] For the figures that follow, see Cairncross, *Costing the Earth*.

consumer market. Investment in pollution abatement and control now accounts for 2-3% of total investment throughout the developed world.[39] This is a large market; in the US sales of environmental protection equipment are expected to rise from $50 billion per annum in 1989 to $200 billion per annum by the end of the century;[40] and major industrial sectors, manufacturing items such as catalytic converters and flue gas desulphurization plants, have sprung up to supply it.

Of comparable importance to catering to the green market, however, is the impact on industry of applying the new environmental insights to their own operations. Perhaps the most compelling example of this is the large chemical companies. Conscious of their low environmental reputation and of the fact that their operations produce large quantities of waste, and particularly toxic waste (chemical companies in industrial countries typically produce 50-70% of all hazardous waste through their operations and products), companies like Dupont, Monsanto and ICI have now launched major environmental strategies to cut their emissions and clean up their processes. These strategies are often significantly more stringent than local environmental laws, both at home and overseas. BP applies Western standards to its operations worldwide, and Monsanto and Dupont have both undertaken to move towards zero emissions. The motivations for this approach vary. It anticipates, probably rightly, a steady toughening of emission standards and helps avoid future bills from retrospective standards such as those imposed by American's superfund legislation (which, for example, has produced a bill for Shell of $1.9 billion for a single, large, case of pollution). It helps to avoid the civil lawsuit or industrial accident which even if resulting from technically legal operations can be deeply damaging to a company's reputation and sales. Such accidents can also be horribly expensive. The largest to date, the *Exxon Valdez*, cost Exxon $2 billion. It can pay in other ways too. 3M Company claim to have saved more than $500 million from their 'Pollution: Prevention Pays' programme introduced in 1975.[41]

This does not mean that business has turned totally green. The environment editor of the *Economist* has persuasively observed that 'since companies are

---

[39] OECD, *The State of the Environment 1991*, OECD Paris 1991.
[40] Tolba et al. *The World Environment*.
[41] Schmidheiny, *Changing Course*.

not altruists, most will only be as green as governments compel them to be. They will do what is required of them and what they perceive to be in their self interest.'[42] Even as popular environmental concern rocketed in the late 1980s, European automobile companies were prepared to be seen lobbying against tough emission controls and US energy companies against carbon dioxide emission limits. One can perceive a distinction between environmental 'slow-track' and 'fast-track' companies. The former tend to be middle-sized, nationally bound firms serving relatively environmentally insensitive publics, and their greenness has often been of the cosmetic, or at most precautionary, kind. Fast-track industry, on the other hand, tends to be more multinational, is often exposed to the best international practice, and has to sell itself to environmentally aware and litigious consumers in countries such as the US and Germany. But there are limits to how far even fast-track companies are prepared to go. Most major companies, for example, have declined to sign up to the 'Valdez principles' – a set of stringent guidelines for environmentally conscious management drawn up by an influential group of US institutional investors following the *Exxon Valdez* incident.

Nevertheless, with the rise of popular environmentalism and major environmental markets, the definition of many companies' 'self-interest' as it touches on the environment has clearly changed. Nor should this change be put down entirely to concern for the bottom line. Managers are citizens and sharers of the public mood too. The consequent emergence of substantial industrial sectors with a major stake in environmental protection has already produced the new, and potentially far-reaching, phenomenon of industries lobbying government on behalf of *more* stringent environmental action, and this will have international consequences. We have already seen a number of cases where governments, having regulated their domestic industry in a certain way, have pressed for that form of regulation to be extended internationally. The ozone layer case is the first example of an entire international industry, having recognized that one sort of market (for CFCs) was on its way out, pressing for a ban, partly in order to guarantee the emergence of a market for the substitutes it was intending to produce.

---

[42] Cairncross, *Costing the Earth*.

6.7 Tropical Forests

If the world cooperated relatively well to deal with the ozone layer problem at this time, it has done less well on tropical forests. Forest destruction, of course, goes back a long way. Its contribution to environmental degradation in ancient times in the Mediterranean basin, the near east and China is well documented.[43] On the other hand, Western growth and prosperity have largely been built on the cultivation of land in Europe and North America which was once virgin forest. In recent times, however, the area of temperate forest has been expanding, and it is in the tropics that deforestation has been concentrated. It is currently estimated that the world's tropical forests are being felled at the rate of about 1% a year, with the destruction particularly concentrated in Central America, west Africa and south-east Asia.[44] This began to cause alarm in the West in the early 1980s, when it was increasingly pointed out that more was at risk than trees. At the local level, forests stabilize rainfall and river patterns, anchor the soil and provide fuelwood and other products, as well as a habitat for 140 million forest dwellers worldwide. At the global level tropical forests are home to perhaps half of the world's living species (most of them uncatalogued, but an important reservoir of genetic material for drugs and other uses) and forest burning contributes somewhere around 20% (the precise figure is highly uncertain) of global carbon dioxide emissions.

Given this combination of interests, one would have expected concern about the tropical forests to be concentrated in the countries, or regions, where the forests stand and where the changes in rainfall and watershed patterns, soil erosion, tribal displacement and other consequences of their destruction would be most acutely felt. This did not happen. Certain developing countries, such as India and Thailand, did indeed pass forest conservation legislation, but failed to enforce it effectively. Other countries, such as Brazil, actively encouraged forest clearance as part of their economic development strategy, and in the 1960s and 1970s built a major road system specifically intended to

---

[43] Ponting, *Green History of the World*.
[44] There is a lot of argument about rates of deforestation, with estimates ranging up to 1.8% per annum and claims that there has been a sharp recent acceleration. *World Resources 1992-93* and Thomas, *The Environment in International Relations*, summarize the most recent estimates.

get new population into the Amazon area. Others again discovered in forest clearance a significant source of foreign income, either through export of tropical hardwoods (as in the case of Indonesia and Malaysia) or through export of cattle reared on cleared forest land (in much of Central America). In African countries especially much deforestation was driven by the demand of impoverished local populations for fuelwood or agricultural land, which no Government was likely to impede.

Thus it was in the developed world that concern was concentrated. This is surprising given that the developed countries have just three direct material interests in the tropical forests. The first two of these, biodiversity and climate change, remain extremely difficult to quantify and were in any case the subject of major separate international negotiations under way at this time. The third, Western imports of tropical hardwoods, was exactly the sort of trade that those most aroused by the issue were determined to stop. It is therefore difficult to understand the high-level Western governmental agitation on this issue at this time, other than as a response to popular environmentalism. Thus the forests provide a rather compelling demonstration of the difference in environmental sensitivity between the developed and the developing countries, with the environmental consequences of their destruction likely to be concentrated in the South but the impulse for their preservation coming primarily from the North.

This raises the interesting question of why Northern concern has been so intense. There are, of course, the biodiversity and climate linkages (although these do not seem to have produced the same enthusiasm for action closer to home, such as higher aid budgets or fuel taxes). It has been plausibly suggested, at least in the case of Amazonia, that a potent contributory factor has been the Western image of the Amazon and its native inhabitants as 'the embodiment of nature'; an image of the region based on a 'literature of denunciation' which goes back 30 years and which has consistently criticized the intrusion of modernity into this ideal realm.[45] In other words, much of the Western perception of developments in the rainforest, including the pace of its disappearance, is an almost mythological construction reinforced by me-

---

[45] David Cleary, *The Greening of the Amazon*, in D. Goodman and M. Redcliff eds, *Environment and Development in Latin America: The Politics of Sustainability*, Manchester University Press, Manchester 1991.

dia repetition; not entirely false, but distant in crucial ways from the reality. It is interesting to reflect on how other popular simplifications and idealizations of complex environmental realities ('acid rain', the 'ozone hole', 'radioactive waste') have driven public policy.

Be that as it may, Western public interest in the rainforests surfaced early in the 1980s and gave rise to steadily growing quantities of Western aid for forest projects, with the total standing at about $500 million per annum by 1985.[46] Over the same period two international institutions intended to help remedy the problem came into existence. The first of these was the International Tropical Timber Organization (ITTO), which began operations in 1987. This is essentially a commodity organization, established to administer the world commodity agreement for tropical timber. Its 48 members are split evenly between the major developing country exporters and major developed country importers of tropical timber. Uniquely among such commodity organizations, maintenance of the 'ecological balance' is among the founding aims of the ITTO and it has indeed set itself 'Target 2000' – that all trade in tropical timber will be on a sustainable basis (i.e. trees will be planted as fast as they are cut) by the end of the century. Moreover, in a manner highly untypical of international organizations, it has implied sharp criticism of one of its key members by calling on Malaysia in 1990 to cut logging by half in the state of Sarawak (which is the source of over 50% of world tropical timber exports and is rapidly depleting its primary forests), a recommendation which Malaysia has accepted.[47]

The other international arrangement concerned with tropical forests is the Tropical Forestry Action Plan (TFAP) which is run under the aegis of the FAO with the assistance of three other organizations. The TFAP started operation in 1987 and its aim is to coordinate aid flows in the tropical forests sector with a view to slowing deforestation. Under the guidance of the TFAP annual aid in the sector had risen to $1 billion by 1990, with 44 aid agencies and over 80 developing countries involved at some point in drawing up the national forestry action plans which are a requirement for benefiting from the TFAP.

---

[46] Thomas, *The Environment in International Relations*.
[47] Porter and Brown, *Global Environmental Politics*.

Despite these impressive statistics, it had become clear by the end of the 1980s that neither the ITTO nor the TFAP was the solution to the problem of tropical deforestation. The commitment of timber-exporting nations to the ITTO's 'Target 2000' is rather belied by the fact that at present less than 1% of tropical rainforest is under sustainable management; the rate of extraction in Sarawak is two to three times the rate of natural regrowth.[48] The TFAP has been accused of being a 'loggers' charter'; it has been estimated that one-third of the aid it distributes goes into commercial logging projects and virtually none of it to tackle the basic causes of deforestation. The plan has been subjected to an intense round of criticism and review by the donor countries, one of the products of which was the emergence of the idea for a world forests convention (of whose fate we will see more below). Both organizations in fact found themselves caught in the contradiction between the Northern definition of a 'forest project' as one which helps to keep trees standing, and the Southern definition, which is often of a plan to cut trees down to assist economic development. Perhaps the most damning criticism of the achievements of the ITTO and TFAP, at least from the donors' point of view, is that there is no evidence that they have brought down the rate of tropical deforestation. Indeed, the most recent evidence is that the rate has accelerated dramatically.[49]

It was against this background that the resurgent Western popular environmentalism of the late 1980s took up the tropical forests as one of its main causes. The number of articles on the tropical forests in the main US newspapers more than tripled from the 1987-8 biennium to 1989-90, the vast majority of them devoted to Brazil, which is host to one-third of the world's rainforest and accounts for 40% of global deforestation. The period saw regular denunciations of Brazilian policy by prominent public figures on both sides of the Atlantic. A stream of US Senators, together with more than 70,000 other visitors a year (up from 12,000 in 1987) found their way to Amazonia. Kayapo Indian chiefs were interviewed on British television chat shows and the Brazilian embassy in London had a more or less permanent antideforestation demonstration sitting on its doorstep. Pop stars staged concerts with titles like 'Don't Bungle the Jungle' and bought chunks of rainforest with the proceeds.

---

[48] Tolba et al., *The World Environment*.
[49] *World Resources 1992-93*.

The climactic event, which gave the campaign a symbol on which to focus, and vastly boosted the media attention devoted to it, was the assassination on 22 December 1988 of the well-known Brazilian forest campaigner Chico Mendes. Within a few months a deluge of books about Mendes in particular and the rainforest issue in general were rolling off the presses. Children in US high schools were collecting pennies to save the rainforest and up to ten Hollywood movies on rainforest issues were reported to be in preparation.[50]

The initial Brazilian reaction to all this Western attention was not welcoming.[51] This is understandable. Two-thirds of Brazil is forest, an area larger than the whole of Europe (excluding Russia). In proportion to its size, the forest contributes little to the Brazilian economy and at the present rate of clearance will last for more than 300 years. Meanwhile Brazil faces major problems with its development plans, not least because of its large overseas debt. Thus President Sarney spoke of 'an insidious, cruel and untruthful campaign' against Brazil. His foreign minister said that 'Brazil does not want to transform itself into an ecological reserve for humanity. Our greatest duty is with our economic development.' Other senior officials referred to 'an economic war disguised in terms of a noble quest for ecological protection. [Westerners] now say that Brazil must give up its sovereignty in Amazonia so that it may be preserved as a kind of Garden of Eden.' They also charged the West with hypocrisy, having grown rich by devastating its own environment and contributed more to global pollution than the vilified Brazil.

Nevertheless, the intensity of the Western campaign caused significant problems for Brazil. It complicated Brazil's access to international aid, and in particular undermined a proposed $500 million World Bank loan to the Brazilian energy sector. It also hindered Brazilian efforts in 1990 to begin to open itself up economically and politically to the West. As a result Brazil gradually bent to the gale of Western criticism.[52] President Sarney began the process of abolishing government subsidies for deforestation. His successor, President Collor, went further, bringing well-known environmentalists into

---

[50] *USA Today*, 24 August 1990; *New York Times Book Review*, 20 May 1990; *Los Angeles Times*, 25 April 1990; *Boston Globe*, 25 November 1990.
[51] For details on the Brazilian case see A. Hurrell, 'Brazil and Amazonian Deforestation', in Hurrell and Kingsbury, *International Politics of the Environment*.
[52] Ibid.

his government and launching a major campaign to slow the pace of deforestation (and bring the murderers of Mendes to justice). This approach was made easier for him by the recognition in certain Western countries that pure criticism of Brazil had proved in some ways counterproductive and that help was needed as well as criticism. Thus in 1989 the UK, followed by others, began discussing with Brazil aid projects to help tackle Amazon deforestation; these discussions have already produced commitments of $250 million for a project to preserve the Amazonian rainforest jointly managed by the Brazilian government, the European Community and the World Bank.[53]

The Brazilian episode nicely illustrates a number of points. First, it underlines the intensity of Western public concern about the environment at the end of the 1980s. It is striking that major Western governments should allow an environmental issue in which they had limited direct material interest to dominate their relations with South America's most important country. Second, it revealed the real difficulties of any international action to tackle the forests issue. Unlike the ozone layer, forests lie within national borders. The reasons for their destruction are often closely bound up with the social and economic problems of the nations within which they stand, being affected, for example by rural poverty, population growth and landlessness. As the Brazilian case shows, it is in any case very difficult for developing country governments to concede any international oversight over what they, and their people, regard as a sovereign national resource, and that difficulty is compounded if the outcome of international involvement is likely to be demands either for policies that will inhibit economic growth or for really profound social changes (such as are now emerging from those who link the problems of the Amazon rainforest with the absence of land reform in Brazil). An additional problem in this particular case were the 'rights' of the indigenous peoples of the Amazon, emotionally espoused by many Western environmentalists but viewed much more dispassionately by the Brazilian authorities who see their role as creating a modern state, and by the Brazilian military who see such particularist assertions as a potential threat to Brazilian sovereignty in the Amazon. Moreover, as so often in the developing world, the Brazilian authorities themselves have very limited control over events in Amazonia. The area is the Brazilian

---

[53] H. French, *After the Earth Summit – The Future of Environmental Governance*, Worldwatch Institute, Washington DC 1992.

'wild west', dominated by local landowners and politicians, where the federal authorities have only limited capacity to stop environmental destruction, however illegal.

In the case of Brazil a variety of special circumstances, notably the unprecedented intensity of Western public interest (which of course fostered governmental generosity when the issue turned financial) and the desire of the Brazilian government to strengthen links with the West for other reasons, enabled these problems to be overcome, at least for the present. But that may not be the general pattern. Certainly a smaller but real upsurge in Western concern about logging in Sarawak in 1991 produced little beyond the deporting of a few Western environmentalists from Malaysia (whose exports of timber and timber products earn more than $2 billion per annum) and the election in Sarawak in 1992 of a provincial government dedicated to maintaining the pace of logging so as to finance industrialization. Meanwhile calls in the West (for example, from the European Parliament in 1989) for bans on imports of non-sustainably forested timber have been met by the timber-exporting nations with (accurate) claims that such action would breach world trading rules. Rather more hopefully, the Dutch have now established sustainable forestry agreements with Costa Rica, Bhutan and Nepal. But even if pursued, such policies would affect only that 20% or so of deforestation which results from commercial logging. Thus at the end of the 1980s the tropical forest problem stood high on the West's political agenda, but with no very obvious solution to hand.

6.8 Moving up the international agenda

The 1980s, like the 1960s, came to an end with a widespread feeling within the West that specific international negotiations to tackle specific issues were no longer enough. The issues were themselves growing vaster. The negotiators had already moved on from acid rain to the ozone layer, and beyond the ozone layer stood the biggest environmental problem of them all – global climate change. Concern within the West that human activity was changing the planet's climate leapt towards the end of the decade. The years 1981, 1983, 1987, 1988, 1989 and 1990 were six of the seven warmest since records began. We have already noted the sequence of extreme climatic events which marked this period. Global climate change was a sufficiently apocalyptic prospect to merit extensive coverage by the Western media, and by late 1988 the airwaves were full of it. Given the intensity of Western public concern, it is unsurprising that the environment forced itself on to all sorts of fora at this time. The period saw a spate of what we have defined as level one responses – agreed but legally non-binding words – to the new international environmental agenda.

The environment played a useful subsidiary role, for example, in the glasnost-inspired *rapprochement* between the USA and the Soviet Union. As the Soviet political system opened up from 1986 onwards, and with accelerating speed as the decade approached its end, so the extent of the environmental degradation within its borders became apparent to its people, to its leaders and to outsiders. Indeed, environmental issues were prominent in the Soviet Union's first ever parliamentary election in March 1989.[54] Moreover, as in the earlier case of acid rain, the Soviet leadership was looking for new and non-contentious subject matter with which to rebuild its relationship with the West and the wider world – particularly given the need for it to demonstrate that it had moved beyond the secretiveness that initially followed the Chernobyl explosion. Hence the Soviet Union launched a number of high-level international initiatives on the environment. In his speech to the UN General Assembly in 1988 Eduard Shevardnadze called for an 'international regime of environmental security' and Gorbachev himself followed this up at the end of the year with proposals for closer international environmental cooperation. The issue was discussed at the 1987 and 1988 Gorbachev-Reagan summits and eventually produced, in 1990, a US-Soviet agreement for joint scientific work on such issues as ozone and climate change. A small strut, perhaps, in the new East-West bridge that was being built, but significant in that it demonstrated that the environment had now taken on such political prominence that it was fit subject matter for discussions between heads of superpower governments.

Other Western heads of government were similarly feeling the need to respond to the environmental alarm of their peoples. The President of France and the Prime Minister of the Netherlands convened in March 1989 at the

---

[54] Feschbach and Friendly, *Ecocide in the USSR*.

Hague a most unusual summit meeting of heads of government of 24 countries, 12 developed and 12 developing, with a view to calling for improved international decision-taking procedures on environmental issues. This was not an entirely happy occasion. The hosts failed to invite the US or the Soviet Union, thus detracting somewhat from its weight, and a number of other countries chose to be represented at less than head of government level (Brazil, in particular, sent only a senior official because of the continuing row about the rainforest). Moreover, the environmental aim of the meeting was, at the insistence of the developing country participants, balanced in the final text by references to the obligations of the industrialized countries to help developing countries. Nevertheless, by the standards of such documents the final declaration of the meeting is strikingly short and clear. It expresses alarm at the 'considerable dangers' posed by ozone depletion and global warming and calls for the establishment of effective international institutions and procedures – even if they have to work by majority vote. Thus the Hague Declaration, which has had little practical follow-up due to the absence of key nations from its list of signatories, underlines the extent of political alarm at the time – even to the point that such fiercely independent political entities as France, Japan, Indonesia and Brazil should be willing to contemplate submitting themselves to international decisions by majority vote on issues with such potentially pervasive impacts on national economic and energy policies as climate change.

France, as current chairman of the group of seven major industrialized nations (G7) followed up the Hague meeting by having the G7 summit in Paris in June that year devote one-third of its communiqué to environmental issues. The G7 had briefly alluded to the environment in its annual statements before, but never at such length or in such detail. In 19 paragraphs the statement ranges over climate, ozone, acid rain, river and marine pollution, desertification, deforestation, nuclear safety, and trade and technology transfer, all in the language of unconstraining evangelism common to such documents – indeed, the only concrete point in the text is the explicit non-acceptance by some members of the Seven (most importantly the US) of the earlier Hague conclusions. Nevertheless, the mere extent of the environmental passage again demonstrated the political need for heads of Western governments to be seen to be discussing the issues.

Nor was this need confined to Western governments. Despite the continued absence of any upsurge in developing countries of popular concern about the environment (and particularly the international environment) comparable to that in the West, the heads of government of the countries of the Non-Aligned Movement (including most of the developing world) meeting in Belgrade in September 1989 devoted three pages of their communiqué to the subject. Their main message was to take up the Brundtland Report credo of sustainable development. Environmental protection in developing countries was an integral part of their development and had to include 'the meeting of the basic needs of all people on our planet... especially a speedier development of developing countries... regulatory regimes must be accompanied by supportive measures to facilitate the adjustment of developing countries... these measures must in particular include net additional financial resources and access to and transfer of alternative clean technologies. A special international fund should be set up for this purpose.' The Indian prime minster had been more specific, demanding the establishment of an $18 billion per annum environment fund and free developing country access to Western technology.[55]

As at the end of the 1960s, then, the stage was set for a major UN conference to address the environmental/developmental issues which had taken on such prominence. The idea of such a conference had indeed already been accepted by the General Assembly at its 1988 session, but it was in 1989 that the modalities were really hammered out. This was not an easy process. In a manner eerily reminiscent of the preparation for Stockholm, a yawning gap instantly became apparent between the environmental preoccupations of the developed North and the developmental ambitions of the South. At the insistence of the latter the conference was (unlike Stockholm) to be entitled the UN Conference on Environment *and Development* (UNCED). After some skirmishing among rival contenders it was agreed that it should take place in Brazil (a site supported by a number of Western countries on the grounds that hosting the conference might compel Brazil to improve its domestic environmental record). The hardest-fought issues in the negotiation of the text of UN General Assembly Resolution 44/228, which was to become the basic text for the preparation of the conference, concerned the passages on aid and

---

[55] *Boston Globe*, 6 September 1989.

technology. At the start of the debate the major developing countries seemed ready to hold the whole idea of the conference hostage unless the resolution guaranteed them satisfaction on these issues. It took difficult and prolonged negotiations, and a lot of encouraging language in the resolution, to get them to hold their aims over for the preparatory process of the conference itself. On aid, the resolution defines the objective of the conference as being 'to identify ways and means to provide new and additional financial resources'; and on technology the objective was 'to examine with a view to recommending effective modalities for favourable access to, and transfer of... technologies..., including on concessional and preferential terms'. These carefully crafted formulae, which were to echo like a mantra through the whole UNCED process, already underline the determination of the South to use the conference to gain better access to Western aid and technology, and the reticence of the North on what they were willing to provide. In a sense the stage was set for a rerun of the New International Economic Order debate in the new environmental setting.

> These carefully crafted formulae, which were to echo like a mantra through the whole UNCED process, already underline the determination of the South to use the conference to gain better access to Western aid and technology, and the reticence of the North on what they were willing to provide. In a sense the stage was set for a rerun of the New International Economic Order debate in the new environmental setting.

At about the same time the UN launched a global negotiation on climate change, and UNEP launched a similar negotiation on the protection of biological diversity. The aim of both these exercises was to have treaties ready for signature at UNCED (the full pattern of the various interlocking pre-UNCED negotiations is shown in Figure 6.1). Resolution 44/228 had set the date of UNCED for June 1992 – two and a half years hence – and had stipulated that representation would be at 'the highest possible level'. This was to be the first ever world summit. In formal terms at least, environment/development issues were now at the very top of the world's political agenda.

Figure 6.1 The structure of the UNCED negotiations

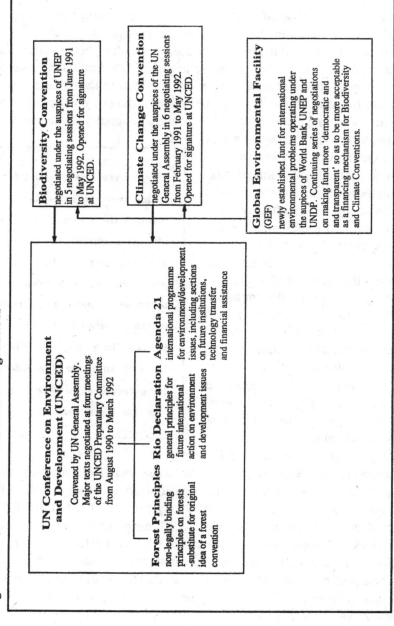

Chapter 7
Climate Change

> *As a rule democracies have very confused or erroneous ideas on external affairs; and generally solve outside questions only for internal reasons.*
> ALEXIS DE TOCQUEVILLE

7.1 The early work

The science of climate change goes back a long way. It was in 1827 that the French mathematician and scientist Jean Baptiste Fourier suggested that the earth's atmosphere traps the heat of the sun in the same way as the glass captures heat in a greenhouse. Moreover, the amount of heat trapped depends upon the proportion in the atmosphere of certain 'greenhouse gases', of which carbon dioxide, methane and the CFCs are the most prominent. Crudely speaking, the higher the proportion of greenhouse gases in the atmosphere the warmer the planet will be, and the lower the cooler. Indeed if there were no greenhouse gases in the atmosphere the earth would be 33°C colder and life might well never have got started.[1] With the coming of the industrial revolution, and the large-scale burning of fossil fuels to provide energy for industrial and home use, the Swedish scientist Arrhenius drew attention in 1896 to the theoretical possibility that such burning could raise the proportion of carbon dioxide in the atmosphere and thus raise the surface temperature of the earth.

This hypothesis attracted some sparse scientific interest over the next three-quarters of a century but only began to earn serious attention with the in-

---

[1] World Meteorological Organisation/ United Nations Environment Programme Intergovernmental Panel in Climate Change, *Climate Change: The IPCC Scientific Assessment*, (ed Houghton et al.), Cambridge University Press, Cambridge 1990.

creased international scientific work on meteorological and atmospheric questions which followed the international geophysical year of 1957-58. In particular, by the late 1960s it had become clear that the proportion of carbon dioxide in the atmosphere was in fact rising and at that time stood about 10% higher than in preindustrial times (the figure now is about 25%). This led to a further intensification of research activity and a degree of political interest, with the Secretary General of the UN referring in his 1970 report on the environment to the danger of a 'catastrophic warming effect'. This point was amplified in the preparatory documents for the Stockholm conference which, as we have seen, agreed on various, more cautious, references to the possibility of human impacts on the climate in its conclusions.

The reasons for this caution are not difficult to discern. The earth's climate is a highly complex physical system, involving not only the entire atmosphere but also the ocean currents, and is in any case subject to wide natural fluctuations of which the best known are the regular ice ages. Our knowledge of the likely impact of human activity on this system was very limited in 1970 and, despite substantial advances, remains so today. In particular, even the best current computer models differ radically in their predictions of climate change at a local level and only very crudely take account of such potentially crucial factors as cloud cover. With only such tools, and in the absence of some compelling shift in weather patterns, it was, and remains, very difficult to identify any change in climate which is demonstrably outside the natural variability of the system and so has to be the result of human factors. Moreover, if in the 1970s the science was vanishingly thin, the politics was impossible. This was a time when Western countries were beginning to take a grip on domestic pollution, but the only subjects receiving serious attention internationally were marine oil pollution and animal conservation. Even at the domestic level, action to protect the environment remained the subject of intensely polemical debate in many developed countries and widespread indifference or even hostility in the developing world. Against such a background any suggestion for world action to protect the climate would have been greeted with, at best, disbelief. It took, as we will see, another 15 years of environmental science and politics even to begin to turn the situation round. In retrospect what is surprising, and perhaps a portent of the ability of the subject to grip the popular and political imagination, is that Stockholm referred to the issue at all, however cautiously.

Nevertheless scientific work continued, most of it under the auspices of UNEP and the World Meteorological Organization (WMO), which organized the first World Climate Conference in Geneva in 1979. This scientific gathering noted that the increased proportion of carbon dioxide in the atmosphere had risen to 15%, put this down 'with some confidence' to (inter alia) fossil fuel burning and deforestation (since burning trees also releases carbon dioxide) and described it as 'plausible' that the outcome could be gradual warming of the lower atmosphere. This slight hardening of the scientific view gave rise to further rounds of international scientific work, data collection and analysis and led to the cluster of meetings which might be described as marking the end of this first, purely scientific, phase of work on climate change. These were the so called 'Villach/Bellagio' workshops which met under WMO/ UNEP auspices in Villach (Austria) and Bellagio (Italy) in 1987. At these meetings, for the first time, a substantial body of the world's climate scientists agreed that global warming was a serious possibility, offered an estimate for its pace (0.3°C per decade), estimated the contributions of the various gases concerned and suggested (in what has been described as an 'audacious piece of international policy entrepreneurship'[2]) an international treaty to cut back the expected rate of release of greenhouse gases, and thus of warming.

7.2 Climate change goes public

At this stage, even after the Villach/Bellagio meetings, the climate issue was still scientific rather than public property. A potent combination of events shifted it abruptly from the one domain to the other in mid-1988. The successful negotiation of the Montreal Protocol in September 1987, and the publicity that it attracted, contributed to already rising Western public awareness of international environmental issues. Perhaps more important, the achievement of the protocol encouraged the view within Western environmental ministries, NGOs and international organizations that the problems of the global atmosphere could successfully be tackled, and so prompted them to place the climate change issue on the international agenda.[3] In particular Canada, which

---

[2] E. Parson, 'Negotiating Climate Cooperation – Learning from Simulations, Theory and History', PhD thesis, Harvard University, May 1992.
[3] *Wall Street Journal*, 17 September 1987.

had been a leading activist on the issue of regulating CFCs, had gained in international standing by hosting the Montreal meeting, and was concerned about US inaction on acid rain, organized an expert conference in June 1988 in Toronto on 'The Changing Atmosphere: Implications for Global Security'. Those invited included scientists, NGOs, international organizations and governments. By later standards the turnout was not particularly high-level. The 300 participants from 48 countries included just two prime ministers (the host and the by now ubiquitous Mrs Brundtland) and seven ministers, with most countries and bodies being represented at relatively junior official level. But the conclusions on climate change (which were of course not binding on Governments) were eye-catching. The conference called for a 20% cut in global carbon dioxide emissions from 1988 levels by 2005 with the eventual aim of a 50% cut, a reduction in deforestation, and the establishment of a 'World Atmosphere Fund' to be financed by a levy on fossil fuel consumption in developed countries. As the first widely endorsed quantitative proposal for action to tackle climate change, these ideas were prominent in international discussion for some time to come.

A great deal of their impact came from the exquisitely precise timing of the conference. The years 1987 and 1988, along with 1981 and 1983, were among the hottest of the century. The end of 1987 had seen a freak hurricane in the English Channel, and a chunk of ice 25 miles by 99 miles broke off the coast of Antarctica. The following year, 1988, brought hurricane Gilbert to the Caribbean and a catastrophic drought to the American midwest. It also brought testimony by James Hansen, a NASA scientist, to a US Senate sub-committee that it was 'time to stop waffling so much and say the evidence is pretty strong that the greenhouse effect is here.'[4] This made an enormous public impact despite the failure of later scientific work to endorse the more dramatic of his assertions.[5]

Suddenly in late 1988 the climate change issue was an urgent item of public business throughout the West. In America two bills were tabled in the Senate on action to tackle climate change, the EPA produced a scary report on the possible impact on the US, and presidential candidate George Bush

---

[4] J. Hansen testimony to US Senate Energy Committee, June 1988.
[5] As summarized in the IPCC reports.

announced that he would use the 'White House effect' to combat 'the Greenhouse effect'. In the European Community the European Commission produced its first 'communication' on climate change, intended to pave the way for subsequent joint community action. Nor was concern entirely confined to the developed world. In September 1988 the South Pacific Forum, which includes the small Pacific island states, many of them low-lying and therefore seriously threatened by the sea-level rises which were expected to accompany global warming, expressed concern and drew attention to the '500,000 environmental refugees' which might be one of the local results of global warming.

7.3 Governmental responses

This eruption of climate change on to the political scene presented most Western Governments with an acute problem. On the one hand, scientific and popular alarm about the phenomenon was genuine; and indeed, early projections of the impact of climate change *were* alarming. The EPA, for example, predicted a shrinkage of US forests, floods along the coastline, major crop losses, increased air pollution and a rise in electricity demand which could require the spending of $110 billion on new power plants over the next two decades.[6] On the other hand, nations which had just agreed with immense difficulty and after six years' negotiation to begin to phase out the $5 billion per year world CFC industry were now confronted with environmentalist demands that they be ready to spend hundreds of billions of dollars on the far vaster task of adapting their economies to use less of one of their most vital inputs – fossil fuels. It is worth underlining this difference in scale between the climate issue and the ozone issue. The CFC industry was of course substantial and linked to other important industrial sectors such as refrigeration and air conditioning. But the scale of the impact of action on ozone is marginal by comparison with action on climate. The latter, if pursued thoroughly, would require total restructuring of an absolutely central industrial sector – energy production – and so would have major repercussions on the whole global economy. In the US alone, the costs of action to cut carbon dioxide emissions by 20% on

---

[6] US EPA, *The Potential Effects of Climate Change on the United States*, Washington, DC 1988.

1990 levels have been estimated at 2.5% of GNP per annum or more,[7] though those estimates were vigorously attacked and later assessments, while remaining highly uncertain, have produced lower figures[8]. Major lobbies, on both sides of the issue, girded themselves for action. In the US, for example, energy interests were organizing in autumn 1988 and mobilizing likely allies in the departments of energy and the interior to discourage any move towards early controls on carbon dioxide emissions.[9]

In such circumstances, the standard response of national governments to domestic pressure is to undertake to study the issue further, and in this case that was the logical reaction at the international level too. Accordingly, in November 1988 35 countries, mostly developed, meeting in Geneva at US instigation but under UNEP/WMO auspices, established the Intergovernmental Panel on Climate Change (IPCC). The purpose of the panel is well summed up by the tasks of its three working groups. Working Group I was to assess available scientific information on climate change; Working Group II was to assess the environmental and socio-economic impact of climate change; and Working Group III was to formulate national and international responses. This was to be no mere coming together of free-flying experts like the Villach and Toronto meetings. The process was intended to be policy-oriented, so the working groups were composed of nationally appointed (and often instructed) delegations. On the other hand, the tasks of Working Groups I and II were essentially to assemble an expert consensus on their areas of study, so that countries tended to appoint their best-qualified scientists as participants. There was early recognition that the process was excessively dominated by developed countries (with a handful of the largest developing countries such as China, India and Brazil also playing important roles), and steps were taken accordingly to encourage more third world participation and to ensure that

---

[7] A. Manne and R. Richels, 'Reducing CO_2 Emissions: The Value of Flexibility in Timing', in B. Flannery and R. Clark, eds, *Climate Change: A Petroleum Industry Perspective*, International Petroleum Industry Environmental Conservation Association (IPIECA), London 1991.

[8] M. Grubb, J. Edmonds, P. ten Brink and M. Morrisson, 'The cost of limiting fossil fuel CO_2 emissions: a survey and analysis', *Annual Review of Energy and the Enviroment*, Vol. 8, 1993; summarises the work on this subject and suggests a realistic range of World GNP lost from halving long run CO_2 emissions (from projected, not current) in the range 0 to 1.5%.

[9] Parson, *Negotiating Climate Cooperation*.

developing country views were given full weight. The target date for the report was late 1990, so as to be ready for the Second World Climate Conference.

While IPCC laboured its way through 1989 and into 1990, individual Western countries were, quite independently of the achievement of the international scientific and policy consensus IPCC was supposed to produce, reaching their own conclusions on how to react to the climate change problem. The Dutch and the Scandinavians, for example, with tacit support from much of the rest of continental Europe, decided that the Toronto approach of targets for carbon dioxide emissions was right. These were countries in whose domestic politics the environment had come to play a major role. Green sensitivity was strong in their electorates. Being small, they were well habituated to the setting of environmental standards through international agreement and had long experience (for example in the North Sea, acid rain and ozone layer negotiations) of setting environmental negotiating targets, which other countries were then pressured by their public opinions into accepting. The Netherlands, as a country particularly threatened by sea-level rise, had an additional motive to press for vigorous action.

> **The Dutch and the Scandinavians .. had long experience of setting environmental negotiating targets, which other countries were then pressured by their public opinions into accepting.**

The issue was more difficult for the larger industrialized countries. The UK and Germany had big fossil fuel industries, and were conscious of the economic impact on them of any cutback of carbon dioxide emissions. France and Japan, by contrast, were already by developed country standards low emitters (France because of its nuclear programme and Japan because of the high energy efficiency of its economy) and were inclined to the view that if restraint was to be imposed, it should first fall elsewhere (for example by imposing per capita rather than gross limits on emissions). Spain was determined to maintain rapid economic growth, which meant increasing emissions from its relatively low starting point. The Soviet Union was just entering the turbulent phase of perestroika and was in no position to make hard economic decisions to help solve global environmental problems.

The situation of the US was particularly ironic. This was the country which had taken a strong international lead on the ozone layer issue (which was seen by many as the most direct precedent for the handling of the climate change problem) and which had directly instigated many of the steps so far taken on climate change, notably the establishment of IPCC. As we have seen, Bush campaigned on the issue on his way to the presidency and James Baker, in his first international speech after being appointed Secretary of State, told Working Group III of IPCC of the need for urgent action on emissions and deforestation even before scientific confirmation was cast iron.[10] But thereafter a reaction set in. In May 1989 the White House blocked US support for the idea of an international climate change treaty at a UNEP meeting and intervened to alter the latest round of the congressional testimony of James Hansen who, as we have seen, was arguing for vigorous action.[11] In the storm of press criticism that followed, the position was reversed – the US accepted the idea of a treaty and offered to host the first negotiating meeting – but the pattern for future US policy was set. It laid strong emphasis on the scientific uncertainties and the vast costs of remedial action. There was no US enthusiasm for policy action over and above so called 'no regrets' measures – policies which, while helping on the climate change front, were also justified for other reasons – such as CFC phase-out, cuts in energy subsidies, and reforestation. In fact, if rigorously applied (a large caveat) such an approach could make a real impact. The US alone is estimated to subsidize its energy industries to the tune of $40 billion per year,[12] while the total cost of the 1990 Clean Air Act amendments is placed at $25 billion per year. But there was to be no US support for what became the badge of climate change respectability, carbon dioxide emissions targets. A number of reasons have been offered for this shift: the lobbying activities of large energy interests, combined with their close links with the Republican Party; the US economic recession and budget deficit, which made expensive environmental commitments difficult for the federal government to contemplate; Republican libertarianism, which

---

[10] *Washington Post*, 31 January 1989.

[11] The amendment was both slight and, arguably, justified. That it provoked such a row underlines the head of steam that was now behind the climate change issue.

[12] I. Brown, *Energy Subsidies in the United States*, Surrey Energy Economics Centre Discussion Paper no. 38, University of Surrey, 1989.

was actively suspicious of environmental regulation (particularly international regulation) and had indeed almost reversed the US position on the ozone layer; and the influence of the President's Chief of Staff, John Sununu, to whom the President reportedly left the issue, and who was himself unconvinced of its scientific merits.[13]

7.4 Pressure on the US

The history of the next year, up to mid-1990, is one of unavailing continental European efforts to push or embarrass the US into changing its position. Occasionally these pressures produced tactical concessions – mostly intended to placate US public opinion – but on the main point, opposition to carbon dioxide emissions targets, the US was immovable. The first encounter of the series was a ministerial meeting on climate change organized by the Dutch as a follow-up to Toronto. This took place in Noordwijk in Holland in November 1989 and was attended by ministers from 72 countries. It took place, too, against a background of continuing high-level public interest in the issue. Mrs Thatcher addressed the UN General Assembly on climate change that autumn. US opinion polls showed public awareness of climate change up to 80% from 60% the year before.[14] Celebrities like Robert Redford were holding 'Sundance Symposia' on the subject.

At the Noordwijk meeting the Dutch and most of the continental Europeans demanded that industrialized countries should set themselves 'as a first step' the target of stabilizing their carbon dioxide emissions by the year 2000. Japan, the UK, the US and the Soviet Union all resisted this. In another swelling strand of the debate the US, UK and Japan also opposed the establishment of a new climate fund to help developing countries tackle climate change, arguing that existing aid institutions should first be mobilized. The upshot was a declaration from the meeting that 'many industrialized countries' wished to see carbon dioxide emissions stabilized by 2000 (which is diplomatic parlance for the fact that a number did not), with references to a 'possible' new international fund. After the meeting the US administration

---

[13] M. Weisskopf, 'The Environment President, Detached and Political', *Washington Post*, 31 October 1992.
[14] *Boston Globe*, 30 November 1989.

(in contrast to the Japanese and the British) again faced intense domestic criticism, to which it responded by announcing (in the President's State of the Union message) $1 billion for climate research, plans to plant a billion trees a year and the US intention to host a 'White House conference' on climate change in spring 1990.

The run-up to that conference was not propitious. In February 1990 half the Nobel Prize winners living in the US, together with half the members of the National Academy of Sciences, appealed for US action to curb greenhouse gas emissions.[15] Then a major and widely publicized row over greenhouse policy blew up within the administration about the text of a speech President Bush was to deliver to the IPCC that month (it is, incidentally, an indication of the political importance attached to the issue by the administration that the President should decide to address such a relatively low-level body as the IPCC). The EPA and State Department argued for stronger US action but were overruled by Sununu. The resulting speech dwelt on the need to undertake more research and not to hamper economic growth, and was characterized as 'limp' by the environmental movement and much of the press. The most direct result of this affair was that Baker withdrew himself from the issue (formally because of his oil interests), leaving the policy field even clearer for Sununu.[16]

Unsurprisingly, the White House conference itself was a public relations disaster. Again President Bush emphasized the scientific uncertainties and economic costs, and his staff did their best to have the conference adopt conclusions to the effect that the first priority in tackling the problem must be more research. Most of the other 16 countries present, particularly the Europeans, forthrightly rejected this approach and spent a large part of the two days telling an avid press that what was needed was US acceptance of carbon dioxide emissions targets.

The next two rounds were played out at a ministerial meeting on sustainable development in Bergen, Norway in May 1990 and at the G7 economic summit in Houston in July 1990. At Bergen it was again noted that 'most' participating countries would stabilize their carbon dioxide emissions by the year 2000. The meeting was chiefly memorable for the acrimony which had

---

[15] *Washington Post*, 2 February 1990.
[16] Weisskopf, *The Environment President*.

change

now begun to enter these exchanges and a near-riot organized by participating NGOs against delegates from countries such as the US and UK whose policies were deemed insufficiently green. Houston was a smaller and more decorous affair but again was accompanied by a build-up of European and press demands that the US change its stance – which it did not.

7.5 Target-setting

Behind all these public pyrotechnics a number of other processes had been under way, leading up to the Second World Climate Conference in November 1990 – which, in view of public interest, was taking on rising political prominence. First, developed countries one by one were completing their internal assessments of what sort of greenhouse gas emissions targets they could take on and announcing the results. In general these assessments came down to more or less public bureaucratic battles between, on the one hand, the environment ministry with an alarmed public opinion at its side and, on the other, finance, industry and energy ministries concerned to minimize the cost and impact on industrial competitiveness of the target chosen. There is no doubt that in a number of countries the issue was significantly complicated by the inbuilt tendency of governments and energy ministries to overestimate future greenhouse gas emissions due to the natural political tendency to overestimate future economic growth, thus making stringent targets in such countries look less achievable than they actually are.[17]

For many countries the straightforward and internationally comfortable solution to these internal arguments was to set their national target at the international 'norm' without analysing in too great a depth at this stage how this might be achieved and what it would cost (a procedure whose precedent was of course the infectious 30% norm of the acid rain negotiations, but which in this case was to lead a number of countries into some subsequent difficulty). The norm around which debate centred was the Noordwijk target of stabilization of carbon dioxide emissions (in general at 1990 levels) by the year 2000. Setting such a stabilization target was made less painful by the fact that virtually all the growth of carbon dioxide emissions in the 1980s in the OECD took place in the US and Japan. Elsewhere, partly as the result of

---

[17] Foley, *The Energy Question*.

the shift we have already seen of heavy industry to developing countries, carbon dioxide emissions were already stable or even (as in the case of the former West Germany) on a pronounced downward trend.[18] Thus, with slight variants, stabilization was the initial target set for themselves by Australia, Canada, Finland, France, Italy, Luxembourg, Norway and Switzerland.

Green electoral pressures and other factors provoked some countries to go further. Germany, a triumphant veteran of the ozone layer and acid rain debates, with a strong green party threat to the ruling coalition and falling carbon dioxide emission levels in any case, announced (to some domestic scepticism) that it would cut carbon dioxide emissions in the old West Germany by 25% by 2005 (emissions in the old East Germany were in any case falling precipitously due to economic restructuring). Similarly, the Netherlands announced that it would stabilize carbon dioxide emissions by 1995 and cut total greenhouses gas emissions by 20-25% by 2000.

The UK concluded that a realistic target was stabilization by 2005 – and instantly faced domestic and European criticism for not setting itself the same target as most of the rest of the European Community. The UK also exposed another fault-line within the West by emphasizing that implementation of its target was dependent upon other countries implementing similar targets. Through the arduous sequence of meetings to come the UK steadily maintained the position that there was no point in Britain or Europe (with 11% of global greenhouse gas emissions) placing an extra burden on its industry to help tackle global climate change if, say, the US (with 18% of emissions) did not do so too. This position was opposed by most other members of the EC, notably by veterans of the 30% club like Germany and the Netherlands who at this stage took the view that Europe should if necessary go ahead alone.

Japan, as usual, was a special case.[19] Public opinion did not make itself felt as strongly there as elsewhere in the developed world, and there was no green

---

[18] In fact over the period 1980-89 CO_2 emissions in the OECD as a whole rose by 0.5% per annum. This is more than accounted for by CO_2 emissions in the USA and Japan (which together constitute more than half of the total) rising by 1% per annum. In the former West Germany they fell by 1.2% per annum over the period. See World Bank, *World Development Report*.

[19] M. Schreurs, 'International Environmental Problems and Japanese Domestic Policy Making', paper presented at the International Environmental Institute Research Semiar Centre for Science and International Affairs, Kennedy School of Government, Harvard November 1992.

threat (or, at that time, any threat) to the political hegemony of the ruling Liberal Democratic Party. But there was widespread political concern to improve Japan's poor international environmental reputation (Japan was at that time being heavily criticized for its massive consumption of tropical timber and drift-net fishing fleet) and there was a significant lobby of Japanese producers of energy-efficient technologies and goods who foresaw a substantial expansion of their markets, domestic and international, in a more carbon-conscious world. It was in this context (aided by some fortuitous juggling with carbon dioxide projections resulting from the wish of the ministry of transport to promote trains rather than automobiles) that a long standoff between Japan's Environment Ministry and its Ministry of International Trade and Industry was resolved by Japan setting itself two targets. It would stabilize carbon dioxide emissions per capita at 1990 levels by 2000 (i.e. a 6% rise over the period) and, if technological development permitted, would in fact stabilize them absolutely (thus linking Japan's target to the international norm).

This national target-setting process communicated itself to one regional body – the European Community. There are a number of reasons why Community member states saw advantage in the Community as a whole pursuing the issue. As we have seen, it was by now in any case a habitual practice for EC member states to cooperate closely on environmental issues, a habit which had just been thoroughly exercised and strengthened in the course of the ozone layer negotiations. A number of small Community member states, notably the Netherlands, hoped that by getting joint Community support for their national policies they would be able to exercise much greater influence on the global negotiations than they would on their own. It is a fact of some importance that it was the environmentally activist Netherlands that chaired Community meetings at climate change negotiating sessions through the whole of 1991. Moreover, as it became clear that the US was going to pursue a line very different from that of most other developed countries, many member states felt that only a united Community position could be of sufficient weight to counterbalance an intransigent US.

The achievement of a common Community position on the central issue of targets was not an easy process.[20] In the Community, as more widely in the

---

[20] Harvard Social Learning Project, 'The European Community and Climate Change', preprint 1992.

developed world, the majority of Northern states had settled on, or beyond, the Noordwijk target of stabilization of carbon dioxide emissions at 1990 levels by the year 2000. But for one major state – the UK – the target date was 2005, and that action was to be contingent on action by other major countries including the US and Japan. The Southern member states, in particular Spain, were quite clear that they expected their carbon dioxide emissions were going to have to continue to rise in the future to accommodate their economic growth. Thus the Community offered a small replica of the shape of the world negotiation. In the view of those member states, such as the Netherlands, who had adopted the international norm there was one major recalcitrant (the UK) which had to be pressured and embarrassed into accepting the norm, and a group of relatively underdeveloped states for whom a let-out formula had to be devised.

Despite these differences the Community achieved a common position through a sequence of bruising meetings in the late summer of 1990. Under time pressure from the imminence of the Second World Climate Conference, these finally culminated in agreement at a meeting of Community environment and energy ministers in Luxembourg in late October 1990 that the Community *as a whole* would stabilize its carbon dioxide emissions at 1990 levels by the year 2000. While no explicit share-out of who would emit how much was worked out (and would indeed have been very difficult to negotiate), the assumption behind this outcome was that the extra cuts made by those who had decided to go beyond the Noordwijk formula (notably Germany) would offset rises in emissions from the UK (which maintained its 2005 target) and the Southern Community states. The question of linkage of European action with that of other industrial powers was left ambiguous, with the UK insisting on such linkage while the other member states opposed it. Nevertheless, the outcome, whatever its hidden ambiguities, did provide the Community with a strong united position to take to the Second World Climate Conference and the negotiation which was now clearly looming beyond. Moreover, by including the Southern Community states it meant that the US was now the only OECD country, other than Turkey, which was not subscribing to a carbon dioxide or greenhouse gas emissions target.

7.6 The Intergovernmental Panel on Climate Change (IPCC)

The other major development of the months before the Second World Climate Conference was the completion by the IPCC of the report on the first round of its work. This was intended as the major intellectual input to the conference and it too had a difficult birth.

Working Group I of the IPCC had been charged with advising on the science. Its report was undoubtedly the most influential of the documents produced by the IPCC. It was a vast act of scientific collaboration, involving in various degrees 170 scientists from 25 countries and 12 international workshops, and described itself as 'an authoritative statement of the views of the international scientific community at this time'. Its introductory section, which is probably as much as most policy-makers ever read, is an astonishingly crisp presentation of the science of the climate change issue. Its central assertion was that rising proportions of carbon dioxide and other greenhouse gases in the atmosphere were consistent with an observed small rise in global temperature and, in the absence of controls, would cause temperatures to rise by $0.3°C$ ($\pm 50\%$) per decade – the fastest rate seen in the past 10,000 years – with accompanying rises in sea level. There have been odd dissident grumblings since from scientists who were involved (particularly in America) that these conclusions should have been more shaded, and the so-called carbon dioxide emission 'scenarios' used by the IPCC – which have (mistakenly but with wide political impact) been taken as predictions of exponentially rising emissions for the foreseeable future – have been widely criticized. Nevertheless, the very clarity of Working Group I's conclusions undoubtedly made a powerful political impact. In the UK, John Houghton (the British scientist who chaired Working Group I) was invited to present them to the cabinet, and the UK subsequently arranged for similar presentations in a number of developing countries. The official reaction of the US was somewhat different: it protested to the British Foreign Office when the Working Group I conclusions were published in book form ahead of final approval by the whole IPCC.

Working Groups II and III were less successful. Working Group II, whose task was to assess the impacts of climate change, found widespread uncertainty and disagreement as to what these would be. While there was a general feeling that the consequences of climate change for agriculture, wetlands, forest, coastlines, desertification and so on could be substantial, and that adaptation costs (at least) could be high, the continuing high level of scientific uncertainty, in particular about how climate would change locally, made it impossible to reach conclusions anywhere near as clear or influential as those of Working Group I.

At about the time that IPCC was reporting, an intellectual counter-current was also becoming apparent. This developed not so much among scientists (although a few dissidents did cast doubt on the apparent scientific consensus on the inevitability and impact of climate change, and were much cited by the US government as a result) as among economists. Eminent economists began to argue that the potential impacts of climate change were so slow in coming (most of the projections having focused on changes over 50 years or more) that it would be a misuse of resources, and an impoverishment of future generations, to spend a significant proportion of GNP on cutting carbon dioxide emissions; the money would enrich future generations more by being used for such purposes as economic development projects whose returns over the period would exceed the costs of letting climate change take place and simply adapting to it.[21] There is little sign, however, that this argument has made much impression in the public, and therefore the political, debate. Similarly, suggestions that global warming might actually benefit certain countries (a British official report pointed in this direction for the UK, and independent economic work suggests the same is likely to be true for the USSR, China and Canada[22]) had little visible effect on the debate, at least in the UK and Canada. It is a fact of some significance for the Western handling of environmental problems that Western publics seem to be more responsive to the alarm of scientists than to the cost-benefit calculations of economists (although when the question becomes one of specific taxes for environmental ends, such as higher gasoline taxes in the US and value added tax on domestic electricity in the UK, public support for environmental action predictably diminishes).

---

[21] See e.g. W. Beckerman, 'Global Warming and Economic Action', and R. Cooper, 'US Policy towards the Global Environment', both in Hurrell and Kingsbury, *International Politics of the Environment*.

[22] W. Cline, *Economic Stakes of Global Warming in the Very Long Term*, Institute for International Economics, Washington, DC 1992.

7.7 The developing countries join the debate

The final IPCC Working Group, Working Group III, had the task of looking at possible responses to climate change, and thus became the cockpit for much of the climate change politics of the subsequent 18 months. Although the bulk of the public argument about, and response to, climate change had so far taken place in the developed countries this was plainly a subject with very large implications for the developing countries too, in terms both of their emissions and of the likely impact of global warming upon them.

Developing countries already account for about 45% of global greenhouse gas emissions (combining the various gases in proportion to their warming effect on the atmosphere – that for methane, for example, being significantly higher than that for carbon dioxide). Much of these emissions in fact come from deforestation and in the form of methane from rice paddies. As industrialization proceeds it is expected that the developing country proportion of total emissions will rise to over two-thirds by 2025.[23] In per capita terms, however, developing country emissions are very low by comparison with those of developed countries (the figure for China is about one-tenth that for the US). There is thus evident scope for disagreement between developed and developing countries as to how much the latter should be expected to contribute to any global effort to control emissions, depending on whether you look at the issue from a gross or per capita point of view and in terms of current, future, or historic contributions.

On the impacts side it seems likely that, in general, the effect of climate change on developing countries will be significantly greater (at least as a proportion of GNP) than on developed countries. The biggest consequences of climate change are likely to be its effects on agriculture and sea-level rise. Developing countries have much longer coastline than developed countries, and are often (as in the case of Bangladesh) much more densely populated, and in many cases are already prone to flooding. They also lack resources for building sea defences. Agriculture in developing countries accounts for a relatively high percentage of GNP (about 17%, compared to an OECD average of 3%) and tends to be more traditional and so less capable of adaptation

---

[23] Figures are taken from *World Resources 1990-91* and IPCC Working Group III Report.

to a changing climate. Key developing countries, at official level (there is no evidence of much interest in the issue at popular level), became concerned quite early about the implications of these factors for them. We have already noted alarm in the south Pacific islands about the likelihood that sea-level rise could affect their very existence.[24] Sea-level rise also threatened major population displacements and economic loss in countries such as China and Egypt. China, along with India and others, became concerned about the possible disruption which changes in weather patterns could bring to intensive agriculture. In sub-Saharan Africa and other arid areas there were fears that climate change could accelerate the pace of desertification. On the other hand, a number of developing countries, notably the large industrializing ones – China, India and Brazil – were also conscious of the centrality for their continued economic development of rising energy consumption, and the need to avoid any external restraint on this.

Developing country involvement in the IPCC process and the various other international meetings devoted to climate change began at a low level but, as the significance of the issue became increasingly apparent, rose through 1989 and 1990. Knowledge of the problem spread through developing country involvement in major conferences, such as those at Toronto and Noordwijk, as well as specifically developing country events, such as meetings on climate change in New Delhi in February 1989, a meeting of small island states in the Maldives in January 1990, and a meeting of African states in Nairobi in May 1990. There was in addition simple political and intellectual osmosis. Climate change was so prominent in Western public opinion and politics through 1989 and 1990 that the elites and governments in the major developing countries cannot fail to have been aware of it, and to have begun to reflect on its implications for them. Finally, there was the impact of the ozone layer negotiation. The 1990 London meeting, by fully integrating developing countries in the process and creating a fund to help them, meant that many more developing countries had civil servants and ministers with some knowledge of international environmental issues and with some bureaucratic stake in en-

---

[24] This alarm caused them to organize themselves into AOSIS, the Alliance of Small Island States, through which they played a much more dynamic and influential role in the forthcoming negotiation than is normally possible for such tiny countries.

suring that their countries' views were taken into account. This was particularly true of India, which felt that it had been disadvantaged by entering the ozone layer negotiation process relatively late, and was determined not to make the same mistake with regard to climate change.[25]

By the summer of 1990 a broad developing country view of the issue had by now evolved. They saw the problem as the fault of the industrialized countries, which were responsible for three-quarters of the accumulated greenhouse gases in the atmosphere. It was therefore the responsibility of the West to cut back its emissions sharply, not least because of the serious threat climate change posed for the developing world. The West must, moreover, give financial assistance to developing countries to help them adapt to the changing climate. As for the developing countries themselves, their first priority remained economic growth. Their greenhouse gas emissions per capita were much lower than those of the West and must be allowed to rise. It was for the West to finance, and provide technology for, any change to third world development patterns intended to cut back that rise. A world climate fund (along the lines of the ozone layer fund) was needed as a channel for these transfers from North to South. Behind this united front there were significant tensions. The large industrializing developing countries were emphatic that there could be no agreement that threatened their plans for economic growth and that large amounts of compensation were therefore needed from the West for any global agreement; while the Africans and AOSIS, both of whom felt much more threatened by climate change, were more willing to look for an accommodation with the developed countries so as to get effective action started.

Nevertheless, for the present the united developing country approach caused serious difficulties for the West. There was no disposition among Western countries to accept historic responsibility, and a lot of pressure to use the climate change process as a means of getting developing country undertakings to limit deforestation. Indeed, given the West's internecine problems on greenhouse gas targets it was much easier for it to write long texts on defor-

---

[25] India, moreover, is the home of the Southern NGO – Anil Agarwal's Centre for Science and the Environment – which has done the most influential work on the problem from a Southern point of view, and in particular drew critical attention to a growing emphasis on emissions of methane (much of which comes from Southern agriculture) as opposed to CO_2 (which is principally an industrial, and so 'Northern', gas).

estation than on fossil fuel policy, while developing countries vehemently pointed out that fossil fuel burning produced three times as much carbon dioxide as deforestation and should be given pride of place. All of these difficulties had already emerged towards the end of the preparation of the Working Group III report, and had turned the final meeting of that group into an all-night session producing a wordy but carefully hedged document whose one important concrete recommendation was that the world should start to negotiate a global framework convention on climate change.

By the time the IPCC came to its meeting in Sundsvall in Sweden in August 1990 intending to finalise its report for the second world climate conference, some 75 countries were represented. At Sundsvall the whole range of disagreements on the issue were put on display. The US wanted the final report to be full of references to scientific uncertainty. European countries were demanding strong texts on carbon dioxide emissions targets and deforestation. Developing countries, notably Brazil, Mexico and China, worked to cut back the language on deforestation and developing country action and build it up on Western responsibility and the need for finance and technology. One or two new problems also appeared. Russia, for the first and last time in the process, gave some emphasis to the possibility that climate change might have beneficial as well as damaging effects.[26] Saudi Arabia and the other oil-

> At Sundsvall the whole range of disagreements on the issue were put on display ... it finished at 4 o'colck in the morning, one day late, with most of the delegates having abandoned their chairs to gather on the front podium and shout at each other. This unconventional procedure produced agreement to abandon a lot of carefully worked out, but still hotly contested, text ... The chief practical effect of this was that the scientific assessment and the recommendation for a global convention were passed up to ministers unadulterated.

---

[26] This position is of course consistent with the possibility that Russia might actually gain economically from climate change. Whatever the reason for Russia's decision not to pursue it, it was certainly politic in view of its growing need for help from the West in other ways.

producing states began to place increasing emphasis on the need for caution and doubt about action whose effect would be to cut back fossil fuel consumption.[27] It also began to become obvious that the domestic preoccupations of Russia and the East Europeans were going to make it very difficult for them to make much substantive contribution to international environmental business in the near future. Having started in a very civilized fashion with songs about the future from children's choirs and an address from the prime minister of Sweden, the meeting finally came very close to breakdown. It finished at four o'clock in the morning, one day late, with most of the delegates having abandoned their chairs in the conference hall to gather on the front podium and shout at each other. This unconventional procedure produced agreement to abandon a lot of carefully worked-out, but still hotly contested, text prepared in the course of the week and to remit the working group reports to the Second World Climate Conference with the briefest of covering notes. The chief practical effect of this was that the scientific assessment and the recommendation for a global convention were passed up to ministers unadulterated.

7.8 The Second World Climate Conference

When the Second World Climate Conference had been originally planned it had been principally intended as a gathering of scientists to look at research results to date and to identify avenues for future research. This intention had of course been overtaken by politics. It was now to be a global ministerial meeting to chart the way ahead on international action to deal with climate change. In a sense, by taking on such prominence the meeting had achieved what it could even before it convened in Geneva on 6 November 1990. Faced with the Second World Climate Conference as a deadline most developed countries, and the European Community, had set themselves greenhouse gas

---

[27] There is an irony in the hostility of the oil producers to action on climate change. Without doubt the international body which has done most to cut global CO_2 emissions is OPEC, through its oil price rises in the 1970s and 1980s. Indeed, one can view the swelling debate between the oil producers on the one hand and the Western advocates of carbon taxes on the other as a dispute about who should get the revenue from the higher energy prices that both sides wish to see.

targets, and the IPCC had produced a clear report on the science and a recommendation that a global treaty be negotiated.

Given the very sharp disagreements among the 137 countries represented in Geneva it was highly unlikely they would be able to extend the international consensus any further. The vast majority of countries were represented at ministerial level, but several heads of government, including Mrs Thatcher and King Hussein, were present to underline the continuing political salience of the issue. The most conspicuous break in the pattern of representation was the US decision to send an undersecretary instead of the head of the EPA. The prospect of substantive progress was further limited by the fact that the formal business of the two days was simply for each delegation to make a (short) statement of its approach.

Events bore out these expectations. The national statements were very much in line with interests and positions which had already been frequently reiterated. Perhaps the most interesting moment in the debate was a clash between AOSIS (pressing for swift action) and Saudi Arabia (opposing it). The final ministerial declaration reflected the pattern of earlier such texts. It endorsed the IPCC science conclusions. Developed countries were called upon to take the lead in reducing emissions, with developing countries taking 'appropriate action'. The need for 'adequate and additional' financial resources and the 'best available environmentally sound technologies' was acknowledged with no great clarity as to how and in what quantity they should be supplied. Various mechanisms, including a new fund, were non-committally identified as possibilities. The text devoted one paragraph to the desirability of energy efficiency and two to deforestation, again with no specificity on future action. The most hotly contested paragraph was the one that applauded, and listed, the greenhouse gas targets set by certain developed countries. At US insistence (and following a characteristic late-night drafting meeting chaired by Tolba) the wording of this was amended so that the US was not conspicuous by its absence from the list.

Finally, the declaration called for negotiations on a framework convention on climate change to begin without delay following a formal decision by the UN General Assembly. Two points are noteworthy about this. First, the European Community had pressed for a reference to the need for simultaneous negotiation of protocols (the aim being, on the ozone layer and acid rain pattern, to press ahead early with carbon dioxide, and perhaps deforestation, protocols) but were blocked by the US, Soviet Union, Saudi Arabia, China and others. Second, the negotiation was not to be conducted under the auspices of UNEP, as had been the case with most previous international environmental negotiations. This decision stemmed largely from developing country doubts about UNEP. Many of them had only limited representation in Nairobi and so limited involvement in UNEP's day-to-day business. They were much more strongly represented in New York. Moreover, they tended to see UNEP, particularly after the ozone layer negotiation, as too much of a technical environmental body to run a negotiation which they saw as being heavily political and in which they wanted the wider developmental and equity aspects of the issue to be fully taken into account. So, as befitted the political importance of the subject, the negotiation took place under the aegis of the General Assembly itself.

At the end of the year the General Assembly duly adopted its Resolution 43/212 which formally created the Intergovernmental Negotiating Committee on Climate Change (INC) and instructed it to produce a convention in time for UNCED.

7.9 The convention negotiations: initial immobility

In one of its phases of fending off critical press coverage of its climate change policy, the US government had announced that it would host the first negotiating meeting. Accordingly, delegates from 107 states, with a whole host of observers from international organizations and environmental NGOs, gathered in Chantilly, Virginia in February 1991 for the opening meeting of the climate change INC. It was like the start of an up-market soap opera. The same cast of characters – Reinstein the folksy but shrewd American, Dasgupta the highly articulate Indian, Metalnikov the wise and genial Russian, Sun Lin the impeccable Chinese, Estrada the jovial Argentinian, Ripert the painstaking French chairman of the committee, and a host of others – were to reappear regularly over the next 15 months in a variety of exotic locales, each broadly maintaining his approach and objectives from place to place, but with just enough movement to keep the spectator interested to the end.

The meeting did not open in ideal circumstances. History in Eastern Europe and the Soviet Union was going into overdrive. Iraq was still occupying Kuwait. Despite the fact that 1990 had (again) broken meteorological records as the warmest year ever recorded, climate change was yesterday's news and was slipping beyond the attention span of the Western public. Coverage of the issue in the US press in 1991, while remaining respectable, was about half the extent of that in 1990. Moreover, the closing phases of the IPCC and the Second World Climate Conference had already given delegates a foretaste of how difficult the negotiations were going to be.

The Chantilly meeting amply confirmed that expectation. The US did cause some excitement at the start by announcing that its greenhouse gas emissions in the year 2000 would be below their 1987 level; but it turned out that this was entirely the result of measures already in hand, notably the phase-out of CFCs, which at that time were thought to be making a potent contribution to global warming. US carbon dioxide emissions remained on a strong upward trend.

The European Community then took the lead in trying to push things forward. Community delegations hoped that after rapidly disposing of the preliminary procedural decisions the INC would be able to get down swiftly to discussions of the draft convention. In practice, however, it took the entire two-week meeting simply to decide procedures. The EC pressed for early discussion on 'sources' and 'sinks' of greenhouse gases – 'sinks' being a new codeword for forests, to get around the growing sensitivity of certain developing countries to any direct reference to the subject. The aim was evidently to start work swiftly on targets for carbon dioxide emission limits and deforestation. Evidence of how difficult this was going to be came rapidly when this was blocked not only by the US but also by the USSR and most major developing countries on the grounds that such discussion was premature.

Even this early in the negotiation it was clear in most delegations' minds that the three big substantive issues were what commitments developed countries should make to limit their greenhouse gas emissions, what commitments developing countries should make in this area, and what arrangements, in terms of aid and technology, would be made to help the developing countries. Precise wording on how these questions were to be handled had to be worked out, and the second week of the meeting saw furious argument, first between the North and the South and then within the South, as to what this should be. This seemingly trivial issue revealed deep tensions which ran right through the negotiation. In the first round the South, led by India, demanded that this procedural decision should in effect commit the West to providing the 'new and additional' resources foreshadowed in the language of General Assembly Resolution 44/228. The developed countries firmly opposed this since in their view the resources made available must depend upon what commitments the developing countries were willing to make. Faced with Western implacability the G77 (the developing countries' caucus) spent the last three days of the meeting furiously arguing among themselves. On the one hand, an increasingly isolated India insisted that the precommitment on resources must come before everything else. On the other, the small island states and the Africans insisted that this was not a point on which to block the whole negotiation. The important thing was to get to work on climate change. India agreed to compromise only at the very end of the session, so that delegates departed for their capitals with some procedural decisions taken but no substantive work done at all.

7.10 New ideas, but no movement

In June 1991 delegates gathered again in Geneva, in even greater numbers. Over 120 states were now participating, together with the normal retinue of NGOs and others. Again, early hopes that the meeting could move rapidly to questions of substance were dashed. It took a week to agree on who would chair the INC working groups, with the job eventually shared – for each group – between delegates from one developed country and one developing country. It was clear that since Chantilly a number of delegations had thought through rather carefully what sort of convention they wanted to see. India circulated a draft squarely placing the burden of action and provision of resources on the developed countries. China circulated an interestingly different text – a framework noting the importance of the problem but containing no substantive commitments (very much along the lines of the Vienna Convention on the ozone layer) – and argued that the level of uncertainty remained too high for anything much beyond this, a view that was extensively echoed in plenary debates by the US, the USSR and Saudi Arabia. The small European countries and the EC maintained their insistence that the US must sign up to carbon dioxide emission targets.

Two important new ideas emerged in the course of the session. The first of these stemmed from the old argument about the establishment of a climate fund – on which the developing countries were still insisting. The bulk of the developed countries now offered their alternative: use of the newly created World Bank/UNEP/UNDP Global Environment Facility (GEF) which (in the West's eyes) had the great virtue of pulling together funding for all the major global environmental problems into one place and being subject to the highly competent management of the World Bank. It would also depend on voluntary contributions from developed countries – unlike the proposed climate fund to which, in most developing countries' minds, contributions would be compulsory. This alternative was rejected by the developing countries, not least because of the alleged Western domination of the World Bank, and the two sides dug in for a long confrontation on the question.

The second new idea was initially entitled 'pledge and review'. It originated in the difficulty of identifying any concrete commitment on greenhouse gas emissions that the developing countries could realistically be expected to make, given the great variety of their situations and the absolute political primacy they attached to continued economic growth. The idea was that instead of a quantified target they should commit themselves to a process. They would, as they judged best, establish their own greenhouse gas limitation strategies but would submit those strategies to regular review by other parties to the treaty. The process of review was intended to encourage gradual reinforcement of the strategies, not least through the allocation of funding. This idea, which came independently in papers from the UK and Japan, was vehemently rejected by India and China (both of whom maintained that no external body could judge their domestic policies), as well as by the NGOs who saw a danger that, applied to developed countries, it could get the US off the targets hook. It nevertheless became part of the mainstream of the negotiation.

With all of this intellectual fertility the Geneva session saw no fall in dissension. The industrialized countries remained split between those who did not want targets (the US and USSR) and those who did (the EC, Japan and most of the rest). The G77 was also split, between those who expected only the developed countries to act (India, with a lot of developing country support), those who wanted only a framework convention (China, Saudi Arabia) and those who saw some need for compromise with the West in order to achieve a convention (the small islands, some Africans and, increasingly, Brazil, which wanted a successful convention for signature at UNCED). These fissures persisted and even deepened into the third session of the INC in Nairobi in September 1991. By now the negotiators (who were building up a sort of camaraderie in the face of adversity) were beginning to put together a text for the eventual convention, but only on easy issues such as research cooperation. There was growing awareness that the hard questions of commitments and resources were likely to be settled only in a final package much closer to the Rio deadline.

The pattern was continued when the INC met for its fourth session in Geneva in December 1991. Discussion of the big questions – carbon dioxide targets or not, GEF or climate fund, 'pledge and review' or not – took the form of sniping from fixed positions. The US remained intransigently against, and Europe intransigently for, targets and timetables. From the G77, India and China continued to underline that all responsibility for action lay with the West, while the Africans and small islands endeavoured to persuade them in private to let the negotiation make faster progress. The Saudis and Kuwaitis continued to argue that more or less any action was premature. Text on the easy issues continued to accumulate, but the hard ones looked frozen solid.

Beneath the ice, however, some movement was beginning to take place, at least in the OECD. A number of the smaller OECD countries were beginning to recognize that the national carbon dioxide emission targets they had set themselves at the Second World Climate Conference and before were not achievable. This realization was brought to a head by a confused meeting of OECD environment and development ministers in Paris in November 1991, where all members of the organization (including the US) apparently agreed to an (undated) stabilization of their carbon dioxide emissions. In subsequent months this led to a spate of 'clarifications'. Australia and Canada were stabilizing greenhouse gas emissions (excluding CFCs), not just carbon dioxide. Norway wanted to be able to offset against its domestic target carbon dioxide reductions achieved overseas through Norwegian aid. Ireland wanted to be able to offset domestic reforestation against its target. The US had never accepted any commitment to stabilization at all. This confusion (coupled with the hope in some OECD countries that the resignation of Sununu in December might make the US more flexible) rather softened the hitherto very hard-edged OECD argument on targets and prompted the additional realization that until the OECD got its act together on this issue it was very unlikely to be able to resolve its differences with the developing countries on resources and 'pledge and review'.

The fifth session of the INC in New York in February 1992 was nevertheless the low point of the negotiation. Much of the interest of participants was less in the plenary discussion than in the private meetings where the OECD tried, unsuccessfully, to sort out its position on greenhouse gas targets. Those discussions grew increasingly bitter as the gap between the US and most other OECD countries remained yawningly wide (not least because, even though Sununu had gone, Bush was coming under right-wing pressure in the Republican presidential primaries). The failure to close this gap communicated itself to the wider meeting, contributing to a number of bad-tempered exchanges between developed and developing countries and what looked like a campaign by India and Saudi Arabia to place pressure on the West (or wreck the negotiation) by introducing language which they knew to be unacceptable into already agreed portions of text. This mood of North-South tension was exacerbated by increasingly brutal Northern statements that in the absence of some Southern commitment to action and acceptance of the GEF as the funding channel, there simply would be no funding. The mood of the meeting was not improved by suggestions (subsequently implicitly confirmed by Bush himself[28]) that if no climate change convention was negotiated President Bush would not go to Rio (thus significantly lowering the impact and value of the summit), a threat whose effect was only marginally sweetened by a US announcement of $75 million aid to help curb developing country greenhouse gas emissions. For the first time there was widespread talk in the corridors of the possibility that either there would be no convention ready for Rio, or it would be at most the framework convention for which a minority of delegations had argued throughout. Delegates dispersed in the knowledge that their next meeting (and necessarily the last before Rio) was just two months away.

---

[28] *New York Times*, 25 March 1992.

7.11 The ice breaks: the Convention achieved

Consciousness of the imminence of Rio and the stakes now riding on the successful conclusion of a convention were beginning to produce movement. In much of the developed world, awareness that if no convention was ready for Rio then the negotiation could trail on unproductively for years, like the Law of the Sea Conference, was beginning to take hold. There was little relish among Western politicians for attending a Rio summit shorn of its most significant agreement. In the US, President Bush's staff were acutely conscious of the political price (in an election year) of wrecking Rio. In the developing countries, even among those most resistant to taking action to help tackle climate change, it was recognized that failure to finalize the convention would at the very least postpone the prospect of significant Western aid and technology to help tackle the effects of climate change, expose Southern countries to the consequences of unconstrained Northern greenhouse gas emissions which, if imprecisely known, remained putatively disastrous, and also threaten other developmental benefits which might flow to the South as the result of a successful Rio Summit.

The immediate upshot of these factors was that, following a further unproductive effort within the OECD in April, a great search began for a compromise on the carbon dioxide targets issue. Reportedly, the US worked out language with the UK which was then included in the clean compromise draft of the whole convention which Ripert was preparing for use at the final negotiating session in New York in early May 1992.[29] Ripert also used a private April meeting of key negotiators, and subsequent bilateral contacts, to try to identify generally acceptable compromises on the other issues which had riven the negotiation.

Ripert's final draft was presented to the resumed New York INC session on 1 May. It was in many ways a masterpiece which cleared away a lot of trivial textual disagreements that had cluttered the negotiation up for months. On greenhouse gas emissions the compromise language that it offered was extremely complex and open to interpretation either as a commitment to stabilization or as something less binding (a situation Ripert characterized as

---

[29] Weisskopf, 'The Environmental President'; Parson 'Negotiating Climate Cooperation'.

'constructive ambiguities'[30]). It was thus instantly denounced as inadequate by the NGOs and some of the smaller European states, notably the Netherlands. On the other hand, after some private high-level contacts between the White House and a number of major European capitals, it also became clear that the larger European states and the US could accept it as the best outcome available (albeit with differing interpretations – the Europeans seeing it as containing an obligation, the US not[31]). On other issues it took two further all-night sessions of Ripert's private group of key negotiators to find generally acceptable compromise language, and the final polishing of the text in plenary was significantly complicated by what looked like a last-minute filibuster by a number of oil-producing states.

Nevertheless the job was done. The GEF was accepted as the 'interim' financial mechanism of the convention (with the permanence of this arrangement dependent upon the outcome of the restructuring of the GEF; see below). Developed countries undertook to provide 'new and additional' financial resources to meet the 'agreed full incremental costs' of actions taken by developing countries under the convention. 'Pledge and review' (which, because of the opposition even the name had aroused, had been merged into the innocuous-sounding term 'reporting requirements', which are in any case standard in international conventions) appeared in a very diluted form: countries were required to supply information on their implementation of the convention, and a body was established to look at this information. There is the vaguest possible commitment to 'promote sustainable management... of sinks' (i.e. limit deforestation). The text was put to the plenary of the INC on 9 May and, after Ripert had threatened to put it to the vote in response to a lone Saudi objection, was adopted *nem con*.

As with the Montreal Protocol, it looks as if the end of the main negotiation marked no more than a short pause for breath in the evolution of international action on the issue. Within two weeks of the final meeting of the INC the European Commission (after several postponements caused by months of infighting between its industrial and environmental directorates) had come up with a proposal for a Community-wide tax on energy as a means of getting

---

[30] Weisskopf, 'The Environment President'.
[31] European Community, Presidency Statement on Community Signature of the Climate Change Convention at UNCED, June 1992.

Community carbon dioxide emissions down. In response to intense lobbying by Community energy, and energy-intensive, industries, the proposed tax was made conditional on similar action by the US and Japan (thus in effect accepting the point long made in isolation by the UK that the Community should not risk its competitiveness by going ahead alone to reduce carbon dioxide emissions). Even in this form the sweeping economic and other implications of such a tax were too great for the environment ministers of Community member states to endorse the idea instantly (thus causing a theatrical announcement by the Community Environment Commissioner, Carlo Ripa di Meana, that he would not attend Rio as there was no Community policy to present there).

The climate change convention was opened for signature in Rio, and was signed there by 153 states (the most significant absentees being Saudi Arabia, Iran and Malaysia). Rio also saw a short-lived attempt by a number of smaller European states, evidently modelled on the precedent of the 30% Club, to place extra pressure on the US by putting together a statement which all developed countries except the US would be able to sign, committing themselves to stabilisation of their carbon dioxide emissions. It is clear evidence of the new cost-consciousness that many countries were now bringing to the climate issue, and their consequent awareness that action must ultimately include the US, that the vast majority of the major emitters swiftly rejected this initiative, which was thus stillborn. The final irony came ten months later, in April 1993, when the newly elected President Clinton (after what looks like the traditional internal struggle between the economic and the environmental sides of the administration – but with the latter significantly reinforced by commitments that Clinton had made during his election campaign) announced that the US would reduce emissions of carbon dioxide to 1990 levels by the year 2000.

7.12 Conclusions

The history of the climate change negotiation leaves a number of impressions. First and foremost is that of the power of a united scientific input and a politically compelling deadline. It is difficult to overstate the achievement of the report of IPCC Working Group I in forcing governments to focus on the climate change issue and participate seriously in the negotiation. The Working Group I report seized and held the intellectual high ground from the moment it was published. It threw on to the defensive those countries with minimalist ambitions for the negotiation, and in particular left US assertions of 'scientific uncertainty' looking like mere procrastination. The convention that emerged does not of course measure up to the problem as described in the report, but in the absence of the report it is very difficult to see any substantive convention having emerged at all. Moreover, if this negotiation, on what everyone knew to be a huge issue on which countries' interests were widely divergent, had not been dovetailed to the Rio Summit it is very easy to imagine it trailing on for years, perhaps producing a Vienna-style framework convention on the way but almost certainly dependent upon a very dramatic shift in popular views (such as might be produced by a major climatic catastrophe) before it could reach any substantive result. It was Rio and the potential political costs of a failed summit which forced the key participants to compromise at the eleventh hour. Even with the subsequent US shift on greenhouse gas targets it is hard to imagine how the gap between developed and developing countries would have been bridged in any other circumstances.

The second point is that, unlike the ozone layer and acid rain negotiations, the more environmentally concerned states did *not* finally compel the others to fall into line. That largely reflects the different balance of environmental compulsion and perceived costs in this case. While the US drought and other climatic disorders of the time played something of the role of the ozone hole and forest death in the earlier cases, the perceived costs (both economic and political) of action were very much higher and encouraged the US, in particular to stand firm. This stance was of course simply reinforced when other developed countries began to identify those costs themselves and to adjust their targets accordingly. It was only subsequently, as the result of a domestic political shift and in the relative absence of international pressure, that the US changed its position.

The third point concerns the evolution of the negotiation between the developed and the developing countries. Of course, as noted above, neither side operated as a bloc – both were riven by deep internal disagreements – and it is entirely misleading in a negotiation such as this (which is, after all, intended to tackle a global problem) to judge who has 'won' and who has 'lost'.

Moreover, the commitments imposed by the convention – whether on emissions, 'pledge and review' or finance – are loosely phrased and will require considerable further work and definition before they really begin to bite on national policies. Reading the convention, however, it is difficult not to be struck by the imbalance in these commitments between the developed countries and the developing countries. In particular, while the former make quite broad undertakings to finance actions taken under convention by the latter, very few such actions are identified, even though developing countries are likely to account for two-thirds of global emissions in 30 years' time. It is interesting to reflect on whether the outcome might have been different if the West had not become so obsessed with its internal differences over greenhouse gas targets.

The final point to note about the convention is simply how remarkable it is that it was achieved at all. It concerns an issue about which scientific uncertainties are still real and large. It points in the direction of regulation of an industrial sector (energy) which is absolutely central to all modern economic activity and all plans for future economic development. It was negotiated in a bare 15 months despite the increasingly overt opposition of a small but influential group of oil-producing states. Its commitments are mild, but it is not the bare framework convention that many states were demanding at the start of the negotiation. Also, as subsequent developments in Washington and Europe have already demonstrated, it is the start of a process which will certainly continue, although it is by no means clear how fast.

Chapter 8

Biodiversity

Progress celebrates Pyrrhic victories over nature.

K. KRAUS

8.1 The issue

The other treaty negotiation which was dovetailed into the preparations for the Rio Conference was on the issue of the protection of biological diversity (biodiversity). This negotiation showed striking, and entirely non-coincidental, similarities to that on climate change, despite the fact that the substance of the issue was totally different.

Concern about the extinction of living species goes back a long way; we have already seen that a number of conservation agreements antedated the 1960s boom in environmental consciousness. The Stockholm Conference produced two pages of recommendations on the maintenance of genetic diversity and the 1970s saw a rather more operational instrument in the form of CITES. With the explosion of international environmental law-making in the late 1980s, however, it was increasingly argued that the time had come to move beyond the species by species approach embodied in CITES and earlier agreements. Human actions were causing 'one of the great extinction spasms of geological history'.[1] Although only one and a half million of the perhaps ten million living species on the earth have been properly identified and described, scientific estimates suggest that between 1% and 5% of all species are now being rendered extinct per decade – which (on conservative assump-

---

[1] E. Wilson, *The Diversity of Life*, Belknap (Harvard), Cambridge, MA 1992.

tions) is between 1,000 and 10,000 times the natural rate.[2] The bulk of the destruction is the consequence of the disappearance of the habitats in which the endangered species reside, for example through the drying out of wetlands and forest clearance. Indeed, the tropical forests are so rich in unique species (it is estimated that more than half of all species live on the 6% of the earth's land surface covered by the tropical forests) that the most up-to-date estimates of species destruction are based on the rate of loss of tropical forests alone.[3] While the vast majority of the species being destroyed are not of the politically emotive sort which generated earlier conservation efforts, being insects, fungi and plants rather than elephants, whales and birds, they do have real economic and ecological importance. The value of drugs manufactured from materials found in tropical forest plants has been estimated at $22 billion per year.[4] But, as in the case of climate, the initial source of concern about global biodiversity loss was not economic, but scientific, with influential scientists arguing forcefully that the wholesale destruction of living species constituted an irreversible squandering of the genetic resources of the planet and a disruption of ecological networks which were little understood but upon which human life itself might ultimately depend.[5]

As with climate, however, there was another side to the argument. Put crudely, this boils down to the observation that, at least since the invention of agriculture, rising human prosperity has been accompanied by the destruction of biodiversity, and that this has not noticeably rebounded to mankind's disadvantage. Indeed, it is precisely those countries where nature has been most thoroughly repressed – in Western Europe, for example – which enjoy the high living standards to which the rest of the world aspires. This does not, of course, mean that as the earth's area of natural habitat continues to diminish there will not come a time when the value of conservation exceeds that of

---

[2] World Conservation Monitoring Centre, *Global Biodiversity: Status of the Earth's Living Resources*, Chapman and Hall, London 1992.
[3] Ibid.
[4] C. Thomas, *The Environment in International Relations*, RIIA, London, 1992; N. Myers, *The Primary Source: Tropical Forests and our Future*, Norton, London 1984. It should however also be noted (Swanson and Barbier, *Economics for the Wilds*) that the value of drugs produced greatly exceeds the value of the genetic resources in them, as the drugs companies have added the substantial costs of research, development and distribution.
[5] J. Dimond, *The Third Chimpanzee*, Harper Collins, London 1992.

continuing destruction. But it does raise the question of whether that time has yet come, or is even close. In this context, attempts to put an economic value on conserving biodiversity have not been particularly helpful to the conservationists' argument. The 'use' value of biodiversity (roughly, what people are willing to do to conserve it now) can of course be captured through normal market and political mechanisms – and, even before the subject leapt up the environmental agenda, was increasingly being so captured, notably through the growth of ecotourism and the progress of the World Conservation Movement (which has brought about a quadrupling in the past 20 years of the world's protected area, to 5.9% of the Earth's land surface[6]). Similarly, the 'existence' value of biodiversity (what people will pay simply to know it exists) can be, and is being, captured through, for example, voluntary contributions to environmental and wildlife organizations, which amounted in the US alone to about $300 million in 1990 – much of this sum being spent on conservation projects in developing countries. Finally, there is the much-touted 'option' value (the value of having species available to meet possible future needs in medicine, agriculture, etc.). A number of commentators hold that this is bound to be significant and point to the myriad stories of plants whose value was discovered only when they were on the verge of extinction (such as the 'rosy periwinkle' which helps treat certain cancers). On the other hand, the rate of biotechnological advance and the abundance of species which have not yet been identified, let alone studied, suggest that from mankind's point of view there is still a more than adequate reserve of planetary biodiversity. Indeed, a recent study points to a widespread view among economists that the option value of biodiversity is unlikely to be large and may even be negative.[7]

Whatever the economic arguments, popular and political concern about species destruction has given rise to a steady growth of local and individual efforts to protect biodiversity in the years following Stockholm. The new

---

[6] Ecotourism is by far the largest concrete value that can be put on diversity and now amounts to an estimated 4-22% of tourism receipts of developing countries (Lindberg, *Policies for Maximising Nature Tourism's Ecological and Economic Benefits*, World Resources Institute, Washington, DC 1991).

[7] See, World Conservation Monitoring Centre, *Global Biodiversity*, for an account of the debate.

emphasis was on protecting habitats (and the entire ecology within them) rather than particular species. I have already mentioned the IUCN's World Conservation Strategy, through which over 40 countries have been encouraged to establish national conservation plans. In 1976 UNESCO's 'Man and the Biosphere' programme designated 67 'biosphere reserves', and that number has grown steadily to about 300 now. A variety of multilateral and bilateral aid instruments began to put money into habitat conservation in developing countries. One or two innovative channels for private finance for this purpose were also opened up, of which perhaps the best-known have been the handful of 'debt for nature' swaps in which Western organizations buy developing country debt (which trades at a substantial discount) and then cancel that debt in exchange for action by the developing country concerned to protect its indigenous biodiversity, for example through the establishment of a reserve. The total value of such schemes by 1992 stood at about $250 million.[8]

By the late 1980s, as we have seen, the expected response to a global environmental problem was increasingly a global environmental convention. Individual efforts, however effective, had two deficiencies. At a substantive level they missed the vast areas of the globe where such efforts were not taking place; and at a political level they were no longer an adequate demonstration that Western governments were tackling the problem on the scale or with the seriousness that it required. Against this background the idea of a global biodiversity convention, originally suggested just after Stockholm in 1974,[9] was resurrected by the IUCN, which began work on a draft in 1987. This was swiftly taken up by UNEP in the intense atmosphere of Western popular environmental concern of 1988, and formal international negotiations began in late 1990. It has been suggested that this sudden acceleration of the timetable, despite the fact that scientific and economic analysis of the biodiversity issue remained very incomplete, reflected UNEP pique at the General Assembly takeover of the climate negotiation, and its determination, together with a number of Western environment ministries, to seize the opportunity offered by UNCED to get a biodiversity treaty much earlier than might otherwise have been possible.

---

[8] G. Conway and E. Barbier, *After the Green Revolution*, Earthscan, London 1990.
[9] M. Tolba et al., *The World Environment 1972-1992*, Chapman and Hall, London 1992.

8.2 The negotiation

The North-South divisions which dogged the climate change negotiation were, if anything, even deeper in the case of biodiversity. There were two reasons for this. First, the developing countries were not expected to take the lead in tackling climate change. The negotiation centred on what developed countries would do, with the concrete obligations of developing countries playing a subsidiary role. In the case of biodiversity, by contrast, the developing countries – or some of them – were in the very centre of the frame. Species richness tends to be higher at lower latitudes and, as we have seen, is particularly high in the tropical rainforest. Thus the vast majority of species, and of species under threat, are concentrated in developing countries. Indeed, of 12 identified 'megadiversity' countries, which together have been estimated to hold 70% of the world's species, eleven are developing (Mexico, Colombia, Ecuador, Peru, Brazil, Zaire, Madagascar, China, India, Malaysia and Indonesia – the one developed member of the list being Australia[10]). So if the problem was to be tackled at all, action on the ground would have to be concentrated in developing countries. This led immediately, as in the case of the Brazilian rainforest, to Southern insistence that Western concerns about biodiversity, even if valid, did not justify any dictation to the developing countries about how they should manage their own natural resources. Action in developing countries must, in their view, depend on the supply of Western expertise and, in particular, funding. Thus, whereas in the climate case the spotlight was principally on the question of emissions reductions by developed countries, with the question of the

> **The North/South divisions which dogged the climate change negotiation were, if anything, even deeper in the case of biodiversity. There were two reasons for this .. the developing countries – or some of them – were in the very centre of the frame .. and even before the negotiation began there was a history of disagreement about the allocation of the economic benefits and technological advances derived from Southern biodiversity.**

---

[10] World Conservation Monitoring Centre, *Global Biodiversity*.

funds they were willing to provide for developing countries taking second place, in the biodiversity case the spotlight was focused principally on the funding issue. The sums looked formidable. Estimates of developing country financial needs to establish the most basic biodiversity protection programmes were in the range of from $10-14 billion per annum.[11]

The second reason why North-South differences ran deeper in this case was that, even before the negotiation began, there was a history of disagreement about the allocation of the economic benefits and technological advances derived from Southern biodiversity. Over the years Western companies have drawn heavily on the genetic resources of the third world in order to develop products ranging from new agricultural plant varieties to pharmaceuticals. Southern demands for some of the profit, or free access to some of the products, of this go back a long way – as does Northern resistance to such demands on the grounds that they would dilute the patent rights of the companies concerned as well as their incentive to continue developing new products. The issue was raised fruitlessly at the Stockholm Conference and again in the FAO in the early 1980s.[12] Plainly the launching of a negotiation in which the North, under pressure from environmentally aroused scientific and public opinion, was seeking the cooperation of the South to protect Southern biodiversity was a golden opportunity for the developing countries to pursue the point again. Thus the Latin American and Caribbean countries in a joint statement at Tlatelolco in March 1991, and 41 major developing countries in a statement following a ministerial meeting in Beijing in June 1991, all emphasized the need for any biodiversity agreement to provide for the 'equitable sharing of the benefits' of the exploitation of biodiversity.[13]

The biodiversity negotiation opened in November 1990 with the developed countries united as they were not in the case of climate change, if only because the bulk of any action agreed had to fall on developing countries rather than on the developed countries themselves. The developing world too was more united than in the climate change case (there was no biodiversity equiva-

---

[11] Ibid.
[12] G. Porter and J. Brown, *Global Environment Politics*, Westview, Boulder Co 1991; L. Caldwell, *International Environment Policy, Emergence and Dimensions*, 2nd edn, Duke University Press, Durham, NC and London 1990.
[13] Economic Commission for Latin America and the Caribbean, Tlatelolco platform, March 1991; Beijing ministerial declaration, June 1991.

Biodiversity

lent of the small islands pressing for an agreement at almost any cost), but on a rather different agenda. While the North foresaw a simple arrangement in which the South would take on conservation obligations in exchange for (limited) Northern financial assistance, the South saw an opportunity to redress the (in their view) festering injustice of their genetic resources being exploited for Northern private profit. Brazil, China and India in particular insisted that they would not participate in any convention that did not provide them with access to the biotechnologies necessary to exploit their own genetic resources. On top of this there were arguments precisely parallel to those in the climate change negotiation about what obligations the developing countries would be willing to take on, and what should be the channel for and governance of financial aid, with the main developed countries arguing firmly for the GEF (funded through voluntary contributions) and the developing countries insisting on a special biodiversity fund (funded through compulsory contributions).

The subsequent history of the negotiation through seven wearying sessions at various spots on the globe very much parallels, both sociologically and substantively, the climate change negotiation. As in the case of climate there was a very slow procedural start, and delegations held firmly to their initial positions until the looming proximity of Rio compelled a compromise at the final negotiating session in Nairobi in May 1992 – just ten days before UNCED began. The whole process was a much lower-profile affair than the climate negotiation. Fewer countries participated. Fewer NGOs observed. Public and press attention was significantly less (unsurprisingly, given the amount of other international environmental business that was going on at the same time). The practical effect of this relative absence of public interest was limited. Although it obviously diminished public pressure for Western concessions to enable an agreement to be reached, in practice the environmental content of any agreement was almost entirely dependent on what the developing countries were willing to concede; and as we have seen, public pressure on environmental issues in the South tends to be a much less potent force than in the North.

The compromise which emerged at the extremely bad-tempered and confused final Nairobi meeting was very much a lowest common denominator.[14] All of the conservation obligations, which include developing national con-

---

[14] For a summary, see M. Grubb et al., *The Earth Summit Agreements: A guide and assessment*, RIIA/Earthscan, London 1993.

servation strategies and establishing systems of protected areas, are qualified by phrases like 'as far as possible and as appropriate' and thus have very limited force. The agreement is vague in the extreme about where action is really needed, and in particular does not include the list of biodiversity-rich sites for which France had pressed and which would have given some real precision to global action to tackle biodiversity loss. On the question of the sharing of the benefits and technologies derived from third world biodiversity, which was probably the most intensely fought issue of the whole negotiation, the agreement provides for 'sharing in a fair and equitable way' and 'facilitating [developing country] access' but this is only an aim, and must be on mutually agreed terms, so that again its force looks extremely limited. Finally, the issue of the funding channel is settled in an even more roundabout way than in the climate convention, with a 'financial mechanism' established in the convention and a separate resolution inviting the GEF to operate it on an interim basis.

8.3 The aftermath

The negotiators themselves were conscious of the deficiencies of their work. At the time of final adoption of the text an extraordinary string of declarations distanced nations from one or another part of it. Malaysia condemned the weakness of the technology transfer provisions and the reference, even in an accompanying resolution, to the GEF. France described the provisions of the agreement as 'too few and too vague' and declined to give its formal assent to the final text. The US, too, saw the text as 'seriously flawed', particularly on its treatment of the financing and technology issues.[15]

With striking promptitude the US followed this declaration up, two working days after the end of the negotiation, with a further announcement that it would not sign the convention. Reportedly, one of the major reasons for this decision was concern in US drug and biotechnology companies about the possibility, as they saw it, that the convention might compel them to share the profits and products they derived from Southern species with the countries

---

[15] UNEP, Conference for the Adoption of the Agreed Text of the Convention on Biological Diversity, Nairobi Final Act, 1992.

from which those species came (although it is very difficult to read the convention as compelling this).[16] It has also been suggested that the very obscurity of the negotiation caused senior US officials to overlook the way it was going until it was too late.[17] At the same time the US privately urged other Western governments not to sign the convention either, and indeed key Western countries were at this time anxiously studying what looked like a very defective treaty. On the one hand, the weakness of the conservation obligations in the treaty and the ambiguity of its finance and technology provisions argued against signature. On the other hand, the agreement did constitute a minimal first step towards global action to tackle a recognized environmental threat. In the climate context, non-participation by the US, with its 18% of global emissions, would have been fatal, while for biodiversity the US presence was desirable (for the funding it would bring) but not essential. There was also the serious political question of the impact on the Rio conference, and on the reputation of individual Western countries within that conference, of a decision not to sign. In this context the attractions of lining up with the US, which had already very publicly isolated itself on climate change and postponed to the last moment the decision on whether President Bush would go to Rio at all, were not undilutedly positive. In short, the US had by now so marginalized itself that there was a real cost, even for its closest friends, in being seen to be on the same side.

At Rio, 153 countries signed the convention, including all the major developed countries except the US. There was, however, strikingly little enthusiasm. Both Jacques Delors (for the European Commission) and Tolba (for UNEP) made it clear that they viewed the convention as an absolute minimum. Again, too, there was the ironic postscript. In April 1993, with the Rio Conference well behind it and international attention to the biodiversity issue substantially down, the new Clinton administration announced that it would sign the treaty.

The chief lessons taught by the biodiversity negotiation are very similar to those from the climate negotiation. Biodiversity suffered (partly because of the haste with which the negotiation was launched) from the absence of a

---

[16] *Boston Globe*, 12 June 1992.
[17] E. Gardner, *Negotiating Survival*, Council on Foreign Relations Press, Washington, DC 1992.

trenchant and internationally agreed scientific assessment of the severity of the loss along the lines of the IPCC Working Group I report. There are good reasons for this. The issue is much more diffuse, much less scientifically definable and quantifiable, and less graphic in its likely impacts and costs than is climate change. So while there is undoubtedly a scientific consensus (or as close to a consensus as can be achieved) that there is a problem, there is much less agreement on precisely what its dimensions and implications are. This meant that those who argued for delay until the scientific case was stronger were in a less assailable position. On the other hand, this is another case where the deadline imposed by the Rio Conference did a great deal to ensure there was a treaty ready to be signed there. The firm commitments in the treaty are undoubtedly even thinner than in the case of climate because the negotiating process produced less willingness to compromise than in the climate case. It is difficult not to conclude that this was because the biodiversity negotiation was much more clearly a North-South negotiation, with the North very reticent on the financial and technology transfer obligations it was willing to take on, and the South equally reticent on the conservation obligations it was willing to take on in exchange. Nevertheless there is now a treaty in place and machinery for continuing work and discussion. It remains to be seen whether, as in other cases we have seen, this will build up its own dynamic and over time generate steadily more substantive action to tackle the biodiversity problem.

Chapter 9:

UNCED: The Preparations

> *Hope is a good breakfast,*
> *but it is a bad supper.*
> FRANCIS BACON

Major UN meetings tend to generate a great deal of preparatory activity. The aim, as we have seen in the case of Stockholm, is to have most of the documents for the conference ready and agreed before it starts. Only major points should be left for the principals to settle.

The preparations for the Rio Conference were particularly large and complex. In part that reflected the size of the agenda, which in principle encompassed all environmental and developmental issues; in part it reflected the political importance which almost all nations attached to the conference, with the North looking for positive results on their international environmental concerns, and the South looking for an outcome helpful to their developmental ambitions; and in part it was simply a consequence of the level at which the meeting was to take place. When heads of government meet, their administrations put significant effort into ensuring that the results of the meeting will be politically and substantively satisfactory. This is a key justification for summits. In the case of meetings of seven heads of government (at the G7 summits) or twelve (at European Community summits) these efforts inevitably generate extended and difficult negotiations before the meeting takes place. Rio was to be an unprecedented meeting of over 100 heads of state and government.

9.1 The Global Environment Facility (GEF)

Two strands of preparation were of course the climate change and biodiversity negotiations. These were not formally part of the Rio preparatory process but separate, freestanding negotiations (under UN General Assembly and UNEP auspices respectively). The only link was the intention to have the conventions ready in time to be signed at Rio. A third strand, which also did not form part of the formal preparatory process for Rio (and indeed was not even under UN auspices), but was intimately linked to it, was the creation of the Global Environment Facility (GEF[1]).

The idea for a financial instrument to help tackle global environmental problems (as opposed to the Environment Fund of UNEP, which was mostly used for local or regional problems) arose from follow-up work to the Brundtland Report. The upsurge of Western popular concern about international environmental issues in 1988-89 helped overcome normal western governmental reluctance to propose new international financial instruments, and prompted the French and German governments to table the idea formally at a World Bank/IMF meeting in late 1989. The Fund, which came into existence as a three-year pilot programme in November 1990, was to be jointly administered by the World Bank, UNEP and UNDP. It was confined to helping tackle specific global environmental problems (ozone, climate, biodiversity and international water pollution) and to paying only that proportion of the costs of projects which was associated with the global, as opposed to local, benefits these projects would yield (for example, although it would not pay for the basic cost of building a coal-fired electricity plant, which would be of local benefit, it would pay the difference necessary in order to build a hydroelectric plant instead, and so save carbon dioxide emissions).

Most developed countries saw the GEF as a means of drawing developing countries into helping tackle international environmental problems, as well as avoiding the danger of an inefficient and expensive proliferation of single-issue environmental funds along the lines of that established under the Montreal Protocol. Centring all such activity in the GEF under what the West saw as the highly competent (and donor-dominated) management of the World

---

[1] D. Fairman, *Restructuring the Global Environmental Facility: Issues and Options for GEF Stakeholders*, MIT preprint, 1993.

Bank looked like an economical solution to this problem. Moreover, its creation offered at least a partial response to the developing country demands for extra funding which the preparations for the Rio Conference had already begun to prompt. Within six months of its establishment almost all developed countries were participants in the GEF, and had committed a total of almost $1.5 billion to it. As we have seen, they were also by then insisting that the GEF should be the financing instrument for the climate change and biodiversity conventions.

The GEF did not go down so well with the developing countries. Their primary financial objective in the UNCED process was to gain extra funding for their own economic development (including, where necessary, tackling domestic environmental problems). For the most part, spending on global problems was not high on their scale of priorities. They were also at this time suffering from a general nervousness about funding from the West. There were two strands to this anxiety. First, the newly democratic East Europeans and Russians had appeared on the international scene as significant potential competitors for Western financial help. Second, they were conscious (as at the time of Stockholm) that Western popular environmentalism could easily turn into 'environmental conditionality' which would require aid money to be spent in line with the environmental priorities of the North rather than the developmental priorities of the South. They consequently saw the priority the West was giving to the GEF as portending Western pressure on them to reorder their own domestic objectives in order to give greater weight to global concerns like climate. They also had problems with the World Bank's primacy in managing the new fund. They had often criticized the Bank's lending criteria as being too tight, and its governance and priorities as being dominated by the developed countries (leading, for example, to excessive concern with the repayment of debt) and insufficiently sensitive to local economic difficulties. They argued that if the GEF was to serve as the financial instrument for the two conventions then all the signatories of those conventions should have a say in the fund management (as was the case with the Montreal Protocol fund) not just the donors (as was the case with the GEF). Thus the question of the organization of the GEF became one of the key battlegrounds of the pre- (and indeed post-) Rio process. The developing countries maintained pressure through a sequence of meetings for what came to be called 'increased transparency and democratization', but made little progress on the precise modalities which would achieve this while allowing the donor governments to retain what they saw as adequate control over the way their money was to be spent. It was against the background of this dispute that developing countries grudgingly accepted the GEF as the financial mechanism for the two conventions – only on an interim basis, until it became clear whether the 'transparency and democratization' issue could be satisfactorily resolved.

9.2 UNCED structures and national objectives

While the GEF dispute was fought out more or less privately at a sequence of discreet meetings organized by the World Bank, the major agenda of UNCED was being manhandled into shape at the much larger and more public meetings of the conference preparatory committee.[2] This committee had four extended meetings (in all the normal places – Nairobi in August 1990, Geneva in March 1991 and again in August 1991, and finally in New York in March 1992). It was attended by virtually all UN member states as well as a huge panoply of international organizations, NGOs and other observers. Its meetings generated a large penumbra of national committees and reports, NGO and business inputs, a special sequence of European Community meetings, discussion at ministerial or higher level in other international formats (notably, but not exclusively, G7 summits, Commonwealth heads of government meetings, OECD ministerial meetings and two unprecedented special meetings of G77 ministers specifically to prepare for UNCED). As was the case with the climate change and other negotiations the preparatory committee became a sort of travelling village of diplomats, officials, experts and other observers who became very familiar with one anothers' personalities and positions and whose united aim was a successful UNCED – even if their definitions of 'successful' differed radically. A key figure in the process was Maurice Strong who, 20 years before, had been Secretary General of the Stockholm Conference and now was appointed Secretary General of UNCED.

---

[2] For a history of Rio and its preparatory process see the *Earth Summit Bulletin* produced daily by the NGOs that were present, and available on ECONET.

the preparations

There was early agreement on Strong's proposals for the main documents that Rio was to produce. There were to be two of them, mirroring the Stockholm format of 'principles' and 'action plan'. The first was to be an 'Earth Charter', setting out agreed principles for global environmental/developmental action in the future. The second, entitled 'Agenda 21', was to be an environmental/developmental action plan for the twenty-first century, covering the full range of specific national and international policy issues connected with environment and development including, specifically, issues such as population, deforestation, international aid, and reform of the UN's machinery dealing with these topics.

We have already noted the difference in Northern and Southern approaches to the subject matter of the conference. The concept of sustainable development, with its elegant marriage of economic growth and environmental concern, went some way to paper over the cracks, but as the preparatory process settled down to specifics (after two meetings almost entirely devoted to procedural skirmishing) the fundamental divergence became more and more apparent – particularly in the context of drafting the Earth Charter. The Northern states, broadly speaking, were looking for a set of principles which underlined the need for countries to modify their economic and other policies to take account of environmental constraints. Thus they wished to see the 'polluter pays' and 'precautionary' principles included, as well as to see prominence given to such topics as population control and natural resource conservation. For the South, by contrast, the key principles were those underlining their right to pursue economic development in whatever way they judged best, and the responsi-

> **The fundamental divergence became more and more apparent – particularly in the context of drafting the Earth Charter. The northern states were looking for a set of principles which underlined the need for countries to modify their economic and other policies to take account of environmental constraints ... for the South, by contrast, the key priciples were those underlying their right to pursue economic development in whatever way they judged best.**

bility of the North to assist them with this. Certainly this entailed avoiding counterproductive domestic environmental degradation, but international environmental problems were, in their view, largely caused by the North and it was for the North to take the lead in tackling them. The right of the South to exploit its sovereign natural resources (such as forests) and pursue economic growth through whatever means it thought best was open to adjustment for global environmental ends only to the extent that the North was willing to pay for that adjustment.

There were, of course, differences too within each camp. Among developed countries the division was between, on the one hand, those (such as the US, UK, Germany and Japan) who devoted a relatively low proportion (about 0.2-0.4%) of their GDP to overseas aid; and, on the other hand, those (such as France, the Netherlands and some Scandinavians) who gave a higher national priority to overseas aid (0.6-1% of GDP). The first group saw in the financial needs of Eastern and Central Europe and the forthcoming world recession tight constraints on their ability to increase aid levels. The second group, without necessarily undertaking themselves to pay any more, saw in Rio an opportunity to press for a general rise in Western aid levels to bring them closer to what they were in any case already paying. Among developing countries, those on the way to industrialization (China, India, Malaysia) tended to place more emphasis on the avoidance of unacceptable environmental constraints on their future policies, while the poorer and more ecologically exposed developing countries (notably in sub-Saharan Africa) were more concerned to see increased Western assistance.

Individual countries and regions also had particular axes to grind and domestic lobbies to placate. The Arabs and Africans were keen to see references to the protection of the environment of people under oppression (e.g. in Palestine or South Africa). New Zealand tried to raise the issue of whaling and had to be reminded (following the unfortunate Stockholm precedent) that this was a subject for the IWC. Germany and Egypt, following the ecological destruction of the Gulf War, suggested that the question of war and the environment might be discussed (causing intense concern in a number of NATO defence ministries at the prospect of environmentalists getting into this delicate area). Malaysia pressed for discussion of Antarctica, but was dissuaded by the Antarctic Treaty states. Canada was concerned about migratory fish stocks. The Africans demanded a special fund to help combat desertification.

the preparations

Well-meaning but utterly unenforceable texts on the importance of women, youth and indigenous peoples were tabled by various Western countries (many of them with lobbies to placate). The one group of countries which played a strikingly low-profile role was that composed of the East Europeans and states of the former Soviet Union, whose domestic preoccupations gave them little scope to pursue clear objectives through UNCED.

9.3 Agenda 21 and the Earth Charter

Confronted by this enormous variety of particular priorities, Agenda 21 and the Earth Charter moved in radically different directions. Agenda 21 became the repository for the full range of environmental and developmental concerns that individual countries wished to see reflected in the outcome of the conference. It became an enormous document, over 500 pages long with 40 chapters and 115 programme areas covering topics including poverty, world trade, population, cities, atmospheric pollution, deforestation, desertification, agriculture, biodiversity, marine resources and pollution, waste management, the role of women, NGOs, indigenous people, science and so on. Given this evolution, it evidently could not be the mandatory global action plan for which some had originally hoped. It became instead a sort of vast and unconstraining menu from which countries could pick and choose actions and emphases according to their own priorities.

In principle (and excluding the few operational chapters of Agenda 21, such as those on finance and institutions) this should have made the document easier to negotiate, as countries simply had to introduce language underlining the optionality of courses of action they had no intention of pursuing. In practice this was not entirely so. A central tension dominated the negotiation of the document. On one side, the North's aim was to emphasize shared global responsibility for environmental action while avoiding language which might imply historical blame or carry financial consequences. On the other, the South was determined to underline the environmental damage wrought by the combination of Northern overconsumption and Southern poverty – with the evident corollary that the North should put more of its income into helping the South. Each side had strong views on problems the other side should be doing more to tackle; the North on population and forests, the South on lifestyles and finance.

The Agenda 21 negotiation thus ranged widely but predictably. For example, it saw a repeat of the battle over the allocation of the benefits from Southern biodiversity which was also being fought in the negotiation of the biodiversity convention (and which came here to exactly the same unconstraining conclusions). The oil-producing states conducted a major campaign to eliminate from the atmospheric pollution section any critical references to fossil fuel consumption. The West had little success in its efforts to introduce strong language on population, partly because it was not prepared to concede the 'balancing' language demanded by the South on Northern consumption patterns, and partly because staunch resistance by the Vatican and some Catholic allies compelled it finally to settle for a very carefully worded chapter on 'demographic dynamics' which nowhere even refers to the desirability of reducing population growth.

Initially, the fate of the Earth Charter seemed likely to resemble that of Agenda 21. Over 150 draft principles were tabled for inclusion. These reflected not only the Northern environmental concerns and Southern developmental concerns noted above but also a whole battery of particular (if broad) political issues ranging through population, women, indigenous peoples, poverty, war, oppression and other familiar areas. At the final meeting of the Preparatory Committee a working group of international lawyers laboured fruitlessly over this mélange for two weeks, with each country and bloc clinging firmly to its own preferred text, until at a very late stage the conference chairman, in a remarkable display of negotiating expertise, boiled the whole list down to 27 concise and (on the whole) mutually consistent, if vague, principles.

These struck a very skilful balance between Northern and Southern concerns, incorporating, on the one hand, the need for all states to pay attention to the environment and, on the other, the special needs of developing countries (and responsibilities of developed countries) with regard to global economic development. A great deal of the final product derives directly from language which had been agreed in Stockholm 20 years before. Stockholm's principle 21, for example, is included almost verbatim.[3] There is also, how-

---

[3] It is, however, noteworthy that the change to Principle 21 is a slight *weakening* of its environmental force. The principle now asserts 'the sovereign right to exploit their own resources pursuant to their own environmental *and developmental* policies'.

ever, some new material – notably the inclusion of the precautionary and polluter pays principles, as well as what was to have been Stockholm principle 20 until it was knocked out by Brazil (the need for states to inform and consult their neighbours on activities which may have external environmental effects). Acceptance of this final compromise had more of the flavour of exhaustion than triumph. In achieving it the name of the document had to be changed from 'Earth Charter' to 'Rio Declaration' – an indication of the view of a number of key states (notably Canada, which made the point explicitly at the Rio Conference itself) – that the text was not the list of unambiguous and binding principles for global environmental management that they had been seeking.

While the negotiators laboriously piled up agreed text on most of the Rio agenda, there remained a cluster of hard items which everyone knew they could not resolve before Rio itself. Unsurprisingly, these were the operational items, upon which concrete decisions were being demanded by one party or another. There were five of them: deserts, forests, institutions, technology and finance.

9.4 Desertification and deforestation

The first two of these issues – desertification and deforestation – were politically linked. We have already seen, as applied to Brazil, the extent of Western popular, and so political, concern about the steady reduction of the tropical rainforest. This unsurprisingly translated itself into a demand, first stated at the Houston G7 summit in July 1990, for a global forests convention. The initial aim of the West was that, like the climate and biodiversity conventions, negotiations for this should be launched immediately so as to have a text ready for signature at UNCED. Indeed, at an early stage the FAO was drafting such a text with a view to convening the negotiation. Certain forested developing countries, however, were not so keen. While there were at least potential benefits to them (in terms of aid and technology transfers) in global agreements on climate and biodiversity, the proposal for such an agreement on forests looked too much like a move towards international surveillance of their management of their own forest resources. Malaysia, in particular, firmly opposed the idea of a convention. The chief Malaysian negotiator said just before UNCED: 'The almost obsessional anxiety to have a forest convention is driven by concerns which have nothing to do with forests or trees. Developed countries wish to appease their public opinion and thus get electoral mileage out of forests... Forests are clearly a sovereign resource – not like atmosphere and oceans, which are global commons. We cannot allow forests to be taken up in global forums.'[4]

The argument about whether or not there should be a convention, largely pursued between the US and Malaysia, ran through the first two sessions of the UNCED preparatory committee and eventually produced a compromise: UNCED would approve a 'Declaration of Principles' on forests and it would be for the conference itself to decide whether that declaration would then initiate negotiation of a convention. Discussion of what was formally entitled the 'Non-Legally Binding Authoritative Statement of Principles for a Global Consensus on the Management, Conservation and Sustainable Development of All Types of Forests' then began, and very rapidly ran into the sand. This was, after all, a document which noone wanted. The West wanted a convention, Malaysia and others nothing. A whole mass of textual disagreements, and in particular the decision on whether there should be a convention or not, were then passed on to UNCED.

The desertification issue arose later than the forests issue, in the form of African demands at the third meeting of the preparatory committee for a desertification convention and fund. In this case it was the West that had doubts. The EC and the US had memories of UNEP's failed desertification programme and in any case saw desertification more as a cluster of local environmental issues than as a genuinely global issue. On the other hand, they were conscious of the likelihood of public criticism of Western opposition to international action on desertification, especially in the heated political atmosphere surrounding UNCED. A natural solution seemed to be to look for a package in which developing countries agreed to a deforestation convention and developed countries to a desertification convention. It was with that thought in the mind of at least some developed countries that the desertification issue was remitted to UNCED.

---

[4] Wen Liang Ting, pre-summit briefing, 2 June 1992, quoted in *Facts on File*, June 1992.

9.5 Institutions

The question of international institutional arrangements for the environment was notionally on the Rio agenda from the start of the preparatory process, but in practice was only tackled at the final preparatory meeting and in Rio itself. As at Stockholm, a variety of motives came together here. In the West there was a general feeling of dissatisfaction with the existing UN arrangements. UNEP had, particularly in the case of ozone, demonstrated its capacity to pilot global environmental business to a successful conclusion; but, at least as currently constituted, it remained small, technical and more peripheral to the main political business of the UN than the present prominence of the environment seemed to justify. The big UN agencies – UNDP, FAO, UNESCO, etc. – were busily affirming the enormous significance of the environment for them, and the importance they attached to it in their programmes; but in reality they had not paid a great deal of attention to it until the subject had moved back up national political agendas (the President of the World Bank, for example, had just announced, in words exactly paralleling those of his predecessor at Stockholm 20 years before, the new emphasis the Bank was giving to the environment in its operations). They had also proved incapable of coordinating their efforts effectively, thus precipitating the collapse of the Environment Coordination Board which Stockholm had established.

> **UNCED was being pushed in the direction of institution creation in the absence of ability to agree on concrete action in other areas. This was not an issue on which the preparatory committee split on North/South lines .. but it was plainly going to take the political heat of UNCED to resolve.**

There was little appetite in the West for the creation of yet another expensive body to complicate further the UN's already feud-ridden environmental machinery. The right way (from the West's point of view) to improve the UN's environmental contribution was to strengthen its centrepiece and most successful player, UNEP, and vastly improve coordination among those agencies with a contribution to make, for example by invoking the authority of the Secretary General. But such technocratic steps did not seem likely to make the political impact which was increasingly necessary – particularly given the low operational content now visible in much of the rest of Agenda 21. In other words, UNCED was being pushed in the direction of institution-creation in the absence of ability to agree on concrete action in other areas.

Developing countries faced a slightly different dilemma. Those well on the way to industrialization, notably China, India and Brazil, certainly did not want a new UN institution examining the environmental propriety of their development plans. This group was particularly opposed to the idea of establishing a continuing system of national environmental reporting to the UN (following up the national reports that were being submitted to UNCED) and to proposals that NGOs (which they took to mean Western environmental NGOs) should be given open access to the workings of any new institution. A number of developing countries were as conscious as the developed countries of the danger of simply adding to the UN system's bureaucratic gridlock. On the other hand, there were evident attractions, particularly for the least developed and most environmentally threatened, such as the Africans, in creating a body which would keep the development debate in the public eye and maintain pressure on the developed countries to raise aid levels (an aim shared by those Northern countries such as France which wished to see other OECD countries increase their overseas aid).

Again, this cluster of problems was left for UNCED itself to resolve. This was not an issue on which the preparatory committee split on North-South lines. One group of countries, including the US, UK, Japan and India, argued for the technocratic solution. Another group, including France, the Netherlands, some key Latin Americans and most Africans, argued for a grand new UN body to act as a continuing focus for the environment/development debate after Rio, and also incidentally to permit Rio itself to make a bigger political splash. It was plainly going to take the political heat of UNCED to resolve this difference, so in the midst of a group of uncontentious recommendations about improving coordination among UN agencies and strengthening UNEP, UNCED was invited to decide on whether it wished to establish

the preparations 219

a new UN body – the Commission on Sustainable Development – to monitor and discuss environmental developmental issues on a regular basis after UNCED was over.[5]

9.6 Finance and technology

Finally, and dwarfing in difficulty all of the other issues that came to UNCED, there were the linked problems of the provision by the North to the South of finance and technology. In the present context both had seen their inception in Resolution 44/228, although both of course went back much further. Following the failure of the New International Economic Order negotiations in the early 1980s, UNCED was the first major multilateral opportunity for the developing countries to renew their demands for a significant increase in the assistance given to them by developed countries. The justification for this demand had changed. In the 1980s it had been based on assertions of the type made in the Brandt Report about the economic linkages between the North and the South. The vogue document now was the Brundtland Report, and the emphasis on environmental linkages. The North, it was argued, should help the developing countries both because of the environmental degradation resulting in the South from poverty and underdevelopment, and because of the need for the cooperation of the South in tackling global environmental problems.

In more specific terms, the UNCED process saw pressure from the South on two particular issues. First, they wanted freer access to Western technology on (in the words of Resolution 44/228) 'concessional and preferential terms'. Second, they wanted 'new and additional' financial assistance. As we have seen, the second of these demands quite rapidly turned into a campaign, led by China and Malaysia, for a large 'green fund' to assist developing countries with their environmental problems. Some big numbers quickly got attached to this demand. At the Belgrade Summit, where it was originally launched, the Indian Prime Minister suggested a fund of $18 billion per annum.

---

[5] Jacqueline Roddick kindly let me see her unpublished paper, 'The South and the Creation of the Commission on Sustainable Development' on the tangled politics of this issue.

When the UNCED secretariat was instructed in August 1991 to estimate the cost of financing Agenda 21 it came up with the figure of $125 billion per annum in Northern aid (as well as $500 billion per annum spending by the South itself). There were constant rumours early in 1992, particularly in the context of Strong's contacts with the Japanese government and his involvement of the former Japanese prime minister Takeshita in UNCED preparations, that sums like $60 billion were under consideration. (By way of comparison, global concessional aid through all sources, multilateral and bilateral, currently totals about $60 billion per annum.)

The Western response was for the most part unforthcoming. On the technology point the US took a particularly hard line, precisely as it had done in the biodiversity context. Most of the relevant technology was in private, not government, hands. It would simply undermine the incentive for Western companies to maintain technological innovation if governments were to expropriate their patent rights in the way developing countries seemed to be demanding. Indeed, on this issue, the US launched something of a counter-attack by demanding the inclusion of language in Agenda 21 on the need to observe and enforce intellectual property rights. Unsurprisingly, the whole argument was remitted to Rio for resolution.

On the finance points, the West's first line of defence was the GEF. Indeed, they had originally expected UNCED to devote itself principally to international environmental problems, and it was for the financing issues associated with such problems that the GEF had been specifically designed. To the extent that there really was a 'North/South environmental bargain' to be had, the GEF expressed rather precisely the amount the North was currently willing to pay for Southern environmental cooperation. Once it became clear, however, that the developing countries had larger financial ambitions from the Rio process the West faced a problem. On the one hand money (as usual) was tight. Germany had just expensively reunified, the US and UK were in recession and the US was increasingly exercised internally about its public deficit. On the other hand there was an evident political need, in order both to ensure a successful conference and to avoid domestic political criticism at a time when much Western public attention was focused on the environmental problems of the developing countries, for some new money (however cosmetic) to be put on the table.

the preparations

Individual Western countries, and groupings such as the EC, therefore began looking for aid announcements they could make at Rio, and there was a lot of hope that Japan in particular would come up with something significant. There was no support within the OECD for any new 'green fund', and the tens of billions of new dollars per year being demanded (let alone the secretariat estimate of $125 billion per year) were dismissed as totally unrealistic. A relevant benchmark for aid levels, however, was the UN's long-standing target that developed countries should give 0.7% of their GNP as aid (which, if fully implemented by all developed countries, would more than double global aid levels from $60 billion per annum to approaching $130 billion per annum). Most major developed countries, with the conspicuous exception of the US, had accepted this target in principle, without setting a date when they would achieve it. Only the Netherlands, Denmark, Sweden and France had in fact achieved it, and they saw an opportunity of using UNCED to push up the contributions of others.

But these Western preparations took time. It is never easy to extract new aid money from obdurate treasuries, and they were only likely to agree at all when the preparations for UNCED, and the political pressures associated with them, were approaching their peak. So discussions of finance throughout the preparatory process amounted to a dialogue of the deaf. The developing countries demanded their green fund. The West pointed to the GEF. Those who had met the 0.7% target were keen to get other Western countries to accept language pointing in the same direction. But the big economies – notably Japan, Germany and the US – would commit themselves to nothing. While many developing country negotiators privately accepted that most of the numbers on the table (such as the $125 billion per year figure) were entirely beyond the bounds of credibility they saw no advantage in moderating their demands until it became clear that the West was willing to discuss figures at all. The closing stages of the preparatory process saw a series of confrontations, both within the developed countries and between the developed and developing countries, as various groups under various chairmen tried to sort the matter out. Despite some energetic jetting around the world by Strong to encourage Western governments to be generous if UNCED was not to be a disaster, noone budged an inch. The issue went to Rio.

Chapter 10

UNCED: The Conference

*Princes should never see each other
if they wish to remain friends.*
PHILIPPE DE COMMINES

10.1 Atmospherics

UNCED met in Rio de Janiero, Brazil, from 3 to 14 June 1992. It was the largest gathering of world leaders that has ever taken place. Representatives from 178 countries attended, including 117 heads of state and government. This alone was quite remarkable. Simply deciding, as had been done in Resolution 44/228, that a conference will take place at the 'highest possible level' is no guarantee that many (or any) leaders will actually attend. Events billed as summits (such as UNICEF's 'Summit of the Child' of a year earlier) have taken place with rather few governments actually being represented at head of state or government level. Moreover, the amount of substantive business done at these events is, in general, inversely proportional to the number of countries involved, which is perhaps one reason why noone had seriously tried to mount a 'world summit' before Rio. The success of this gambit in the case of Rio depended crucially on building up a 'critical mass' of influential heads of government publicly committed to attend so as to coax waverers into deciding to attend too, if only because their absence would stand out and might attract domestic criticism. With summits, as in other areas of international politics, it is safer to be part of the crowd. Very early in the chain reaction came Mrs Thatcher, who very swiftly and unambiguously said she would go (a commitment reiterated by Mr Major when he succeeded her).

Most other EC leaders, as well as those regularly at the forefront of international environmental action, the Canadians and Scandinavians, also signed up early. The President of Brazil and a large number of South Americans would inevitably be there, and the inclusion of development in the conference remit created a presumption that many developing country leaders would try to be there. By now the snowball was rolling well, although some very big names – notably the president of France, the prime ministers of India and Malaysia, and of course President Bush joined the list very late.

In addition to the heads and their delegations, 1,400 NGOs (one-third of them from the developing countries) sent observers. In all, the conference had 35,000 accredited participants as well as a steady stream of visiting celebrities ranging from Roger Moore to Pele. Eight thousand journalists covered the event. For more or less all of its ten-day length it was front-page news throughout the developed world and in much of the developing world as well.

There were a variety of accompanying events. 'EcoTech' was an industrial fair at which 400 companies displayed their latest environmental technologies. Indigenous groups from around the world set up a temporary village near the UNCED site to 'stake their claim as the true voices of the earth'.[1] Rio, in the period of the conference, saw over 180 artistic events on the environment and nature including *Midsummer Dream in the Amazon Forest* by Werner Herzog out of Shakespeare.[2] By far the largest accompanying event was the Global Forum, modelled on Stockholm's 'Environment Forum' of 20 years before. This was a huge and colourful gathering of environmental and developmental NGOs, scientific, women's and other groups. It attracted over 18,000 participants and functioned as a combination of street fair, trade show, public meeting and political demonstration. Wares on offer ranged from sun-heated cookers to new age philosophies. In a manner strikingly reminiscent of its Stockholm prototype the forum ran out of money in its first week and had to be rescued by a few of the governments meeting 20 miles away at the official conference.

However, again as at Stockholm, many of the NGOs were in close contact with (or actually members of) their national delegations, were kept fully up to

---

[1] *Boston Globe*, 29 May 1992.
[2] *New York Times*, 1 June 1992.

date with the way the negotiations were developing and were able, both through the (by now traditional) publication of the conference newspaper, and by more direct lobbying and drafting, to influence the conference outcome. NGO involvement in the conference and its preparatory process is noteworthy in two other ways. First, there was a much higher proportion of Southern organizations than had been present at Stockholm 20 years before (although they were still swamped by NGOs from the North). Second, the gap between the positions argued by Northern environmental NGOs and Southern developmental NGOs was almost as wide as that between their respective governments, as became soberingly apparent when they tried to negotiate 'alternative treaties' to those being discussed in the main conference and finally emerged with a very diplomatic 'code of conduct' in which they agreed to respect one another's views and not to seek to force solutions on states for which they might not be appropriate.[3]

Participants at the official conference too had their wares on display, albeit in the more sober form of ministerial and prime ministerial speeches. These reflected the shape of the debate as it had emerged during the preparatory process. Ministers from developed countries tended to dwell on the seriousness of global environmental problems and the need for North-South cooperation to tackle them. Thus the President of Portugal, speaking on behalf of the EC, stressed 'the need for shared responsibility for the global environment, exemplified by the climate change convention'. The Prime Minister of Canada called for the negotiation of a 'real Earth Charter'. The Chancellor of Germany repeated the call for a forests convention and the President of France emphasized global interdependence. The developing countries focused more on their domestic environmental problems, especially those resulting from poverty and underdevelopment, and pressed the case for more generous aid from the North. Thus the Prime Minister of Pakistan, speaking on behalf of the G77, said that 'the cause of the present economic and environmental crisis is an unjust world economic order'. The provision of additional financial resources was 'the essential condition for the implementation of UNCED results'. The Prime Minister of India repeated the call for a green fund. The President of Brazil said that questions of environment and economic develop-

---

[3] M. Grubb et al., *The Earth Summit Agreements: A guide and assessment*, RIIA/Earthscan, London 1993.

ment were inseparable. The Prime Minister of Guyana put the point rather more picturesquely: 'the tree of sustainable development cannot flourish in the infertile soil of poverty.' The Prime Ministers of China and Malaysia made strikingly tough speeches. Both emphasized that the developed countries did most of the polluting, and it was on them that the onus of protecting the global environment must principally fall. They should also provide additional finance and technology to help the South with its more pressing problems of poverty and local environmental degradation. A number of smaller developed countries endorsed Southern demands for more aid. Both the Norwegian and the Swedish Prime Ministers emphasized the need for all developed countries to meet the 0.7% target.

The public speeches also reiterated a lot of other well-established policy positions. The Africans, with widespread developing country support, called for a desertification convention. One or two of the East Europeans, notably Russia, drew attention to the environmental damage done by 50 years of communism and called for access to international funding on the same basis as that offered to developing countries. Oman, Qatar and Saudi Arabia warned against precipitate action (such as energy taxes) to cut carbon dioxide emissions. The Holy See attacked Western attitudes towards population growth in the third world.

A key participant in the public debate was of course the US President. He spoke against a very difficult background. A combination of factors had, in many eyes, made the US the villain of the conference. There was the American refusal to accept greenhouse gas targets, their decision not to sign the biodiversity treaty, their hard line (in common with others) on the availability of new funds for the developing countries, and President Bush's very late decision to attend UNCED at all. Before he arrived there was another major embarrassment in the form of a leaked memorandum from the chief US negotiator, Reilly, revealing sharp dissension within Washington on what should be done about the biodiversity treaty.[4] A US attempt to smooth the way for the President's arrival by announcing an increase of $150 million in US aid

---

[4] Cf. Gardner, *Negotiating Survival*. The split was between, on the one hand, the State Department and the EPA, who were looking for ways to mend fences with the rest of the world and, on the other, Vice President Quayle's 'Competitiveness Council' which saw 'no votes in the environment' and economic costs attached to environmental regulation.

the conference   227

for forests misfired as many countries interpreted this as a sop to conceal inflexible American policy in other areas. The US delegation to the conference was not a happy place. Reilly (in a later leaked memorandum) described his role at Rio as 'like taking a bungee jump while somebody cut his line',[5] and other members of the US delegation spoke of a 'firestorm of anti-American sentiment'.[6] Nor did White House briefings that US allies 'paid lip service' to the environment and that Germany and Japan were engaging in a guilt-induced effort to be 'politically correct' help matters.[7] In the circumstances President Bush's words were interpreted by many as defiant rather than conciliatory when he said: '...it is never easy to stand alone on principle, but sometimes leadership requires that you do. And now is such a time... America's record on environmental protection is second to none... I did not come here to apologize.' This quasi-isolated and embattled atmosphere surrounding the most important Western delegation significantly complicated the closing phases of the conference. There are signs that it encouraged a hard-line attitude in certain developing countries. The Malaysian Prime Minister, for example, said the US position had rendered the climate change convention 'inequitable and meaningless'.

Behind the public debate, work was under way to finalize the remainder of the business of the conference. The biodiversity and climate conventions were opened for signature and received nearly 160 signatures each, that is, nearly 90% of all countries attending, an extraordinarily high number for any treaty within such a short time of negotiation, and even more extraordinary given the wide ranging and contentious subject matter of these two particular treaties. There was also, as we have noted, the abortive effort to produce a declaration on greenhouse gases on the model of the 30% Club; this was eventually diluted into a reaffirmation by the European Community, and some others, of their intention to stick by their previously announced greenhouse gas targets.

The Rio Declaration (in a manner highly reminiscent of the Stockholm Principles 20 years before) had been carefully passed on to the conference as a delicately balanced text which could provoke a major row if tampered with.

---

[5] *New York Times*, 2 Aug 1992.
[6] *Atlanta Journal*, and *Atlanta Constitution*, 7 June 1992.
[7] *Washington Post* and *Los Angeles Times*, 10 June 1992.

One or two countries nevertheless hinted that they might have amendments to propose (the US in particular was reported to be considering holding the declaration hostage for changes it wanted elsewhere in the conclusions, and there was an Israeli-inspired problem with the declaration's reference to 'people under occupation'). Consciousness of the sensitivity of the compromises embodied in the declaration, and a certain amount of arm-twisting, discouraged such thoughts so that at the end it was adopted unchanged.

This left Agenda 21. About 85% of this had been completed at the final preparatory meeting. Now that the moment to find agreed, if empty, verbal formulations had really come, most of the rest was finalized relatively painlessly at official-level meetings in the background of UNCED. The only real difficulty was to find language for the atmospheric pollution chapter which was sufficiently vague on the issue of reducing fossil fuel use to be accepted by certain oil-producing countries.

10.2 The five big issues

There remained the five big open operational questions. In the political hothouse of UNCED, with the watching world awaiting agreed conclusions, two of these were quite rapidly disposed of. The political arguments for being able to display a new institution as part of the UNCED product outweighed bureaucratic caution about whether such an institution was really needed, and it was agreed that the UN Commission on Sustainable Development (CSD) should be created. It would meet regularly to maintain the international dialogue on sustainable development and follow up the conclusions of UNCED. The major Western opponents saw less cost in conceding UNCED a 'success' on this issue than in giving ground on other potentially much more expensive items such as aid levels. Thus the main point of argument at UNCED was the resistance of some large developing countries, led by India and China, to any requirement that they report regularly to the UN on their environmental actions. This was accommodated firstly by focusing the attention of the CSD much more precisely on the financial and technological contributions of the North than on any specific action by the South, and secondly by diluting the reporting requirement to almost total vacuity – 'information provided by governments' – with the details remitted to the UN General Assembly to be sorted out. Similarly, the arguments about technology transfer boiled down to

the conference

finding words which referred approvingly (but non-bindingly) to the desirability of transferring more technology to developing countries, and of maintaining intellectual property rights. Such words were duly found.

Resolution of the arguments about desertification and deforestation was more difficult. An agreed text for the forests declaration only emerged very slowly and painfully as those developing countries particularly emphatic on the inviolability of their sovereignty in this area (notably Malaysia and India) scrutinized every word.[8] In the meantime pressure on the West to agree to a desertification convention grew. African speeches, both in the public debate and in the backstage negotiations, focused on this point with a great deal of support from the rest of the developing world. The US, perhaps concerned about its isolation and unpopularity in other areas, succumbed on this point fairly early, leaving the European Community in isolation trying to hold off a decision until the question of a forests convention was also ready to be settled. This proved impossible. The political price for a number of Community governments in being seen to block an almost unanimous demand for international action (however unlikely to be effective) on third world, and particularly African, desertification was too high. UNCED agreed unanimously on 10 June that there should be a negotiation leading to a desertification convention.

The discussion on forests came to a climax a day later. By now most of the text of the declaration (which sets out a very laudable but, as the title makes clear, non-binding statement of the principles of environmentally conscious forest management) had been settled, and the only major outstanding question was whether it should also call for the negotiation of a forests convention. The West, led by the US (Bush having announced that a binding forests convention was America's principal objective at UNCED), argued hard for this. India and Malaysia were adamantly against. The compromise which kept the door ajar was that the principles should be kept 'under assessment

---

[8] Malaysia did give a new, and elegant, twist to the discussion by proposing a binding forestry agreement which would commit all countries to maintain 30% of forest cover, or fund forests in other countries if they could not. Vast differences of geography make it difficult to see this proposal going very far (what about countries which are both poor and lack forests?) and it was not seriously pursued in the private negotiations. But it did underline a glaring inconsistency in the West's approach, namely that it expected the South to maintain its forests but was not willing to come up with significant funding to help them do so.

for their adequacy with regard to further international cooperation on forest issues'. In this form (and to much criticism by Western NGOs) the declaration was adopted by UNCED.

The outstanding, and hardest, issue was the question of finance. Various Western government efforts to find new aid announcements to make at the conference had borne fruit. Germany, with widespread Western support, called for a tripling of the GEF to about $4 billion. The European Community, by adding together individual national commitments, offered $4 billion to help developing countries with the implementation of Agenda 21. Japan announced that it would increase its environmental aid by about $1.5 billion per year over the next five years (which was significantly less than many had been hoping for).

Still the central question remained whether there would be any real Western commitment on a long-term expansion of aid levels. This issue was argued over in a huge variety of different groups right up to the end of UNCED (and indeed was only finally resolved after most of the heads of government had departed). It was debated in the EC, the G7, the G77 and in various informal 'contact groups' bringing together key developed and underdeveloped countries. Faced with a united developed country refusal to contemplate any new 'green fund', developing countries quite rapidly brought their sights down to getting a firm date by which developed countries would achieve the 0.7% target (the year 2000 was suggested) and a firm commitment to a substantial expansion of the 'soft loan' funds available to the World Bank. As we have seen, these ideas also had some support within the West, but the big spenders (notably Germany, Japan and the US) remained firmly opposed and, as time ran out, the final compromise offered little assurance that aid levels were going to rise. The 0.7% target would be achieved 'as soon as possible' and countries would give 'special consideration' to a call by the World Bank president for a substantial expansion of its soft loan facility. In this unconstraining form the finance chapter, together with the rest of Agenda 21 and the Rio and Forests Declarations, were adopted at the closing session of the conference on 14 June.[9]

---

[9] Subsequent developments have underlined precisely how unconstraining the finance conclusions were, with both the EC and World Bank components producing very little new aid (*Economist*, 12 February 1994).

10.3 The lessons

The products of Rio generated quite a lot of disappointment, both at the time and subsequently. At the closing session Maurice Strong referred to 'agreement without sufficient commitment' and the UN Secretary General said 'one day we will have to do better'. There was widespread NGO criticism of the toothlessness of the conclusions on forests, and developing country unhappiness at their failure to achieve concrete long-term commitments on aid levels.

Certainly, reading through the documents produced by the conference is not an inspiring experience. The North-South tensions which dominated the conference are emphasized to the point of tedium, and the process of compromise required to achieve any outcome at all has rendered most of the language so vague and non-operational that it is hard to imagine it having any direct impact on national policies in years to come. In some cases (e.g. population) even the nature of the problem is not immediately apparent from the texts; and it is hard to know what use international jurisprudence is to make of Principle 25 of the Rio Declaration, which states simply that 'peace, development and environmental protection are interdependent and indivisible'. In other words, Rio produced a swathe of what we have defined as level one responses to the problems it addressed which by a substantial margin fail the 'concreteness test' for the likelihood of their swiftly being converted into level two or level three responses. Of the more tangible products of the conference, it is difficult to be confident that the Sustainable Development Commission will end up as any more than a new UN talking shop with very limited impact on the world outside, or that the negotiation of the desertification convention will be any more than yet another sterile confrontation about Western aid levels, this time in a desertification context.

> **The first lesson of UNCED is the limited ability of even global summits to shift substantial groups of nation states from well entrenched positions ... but as with Stockholm, the apparent flaccidity of the formal conclusions of UNCED are by no means the whole story.**

Thus the first lesson of UNCED is the limited ability of even global summits to shift substantial groups of nation states from well-entrenched positions. There was no way in the economic circumstances of mid-1992 that the major aid donors were going to commit themselves to significant increases in their aid. Nor was there any way that developing countries with significant areas of forest were going to make their use of those forests subject to international legal restraint.

But, as with Stockholm, the apparent flaccidity of the formal conclusions of UNCED is by no means the whole story. Rio had other products. Without the Rio deadline, the biodiversity and climate change conventions would not have been completed – and it is possible to argue that the climate change convention alone justifies all the prime ministers of the world getting together for a few days in Brazil. Moreover, in the case of Stockholm one of the principal consequences of the conference was that it set a number of countries (notably India and China) thinking seriously about environmental issues and adopting policies to tackle them. UNCED was a much larger and more highly political event than Stockholm and may well have pushed this process further forward in a large number of countries.

Stockholm, too, produced an apparently unpromising institutional outcome. For a long time UNEP was a marginal player on the UN and world scenes, but over time it has become an effective midwife to a range of significant environmental agreements, such as those on the ozone layer. While the prospects for the SDC of undergoing a similar metamorphosis must be rated as low (particularly since its central functions simply duplicate those of the UN's enduringly ineffective Economic and Social Council), the other institutional effects of Rio may prove to be of more benefit. The conference has pushed the environment up the agenda of the major UN agencies, notably the development agencies such as the UNDP and the World Bank.[10] No doubt, as before, some of this effect will wear off with time, but it is reasonable to expect (as with the environment in national politics) that some will endure. It has also pushed three key agencies into working together in the GEF, potentially a

---

[10] In the run-up to and aftermath of Rio, for example, the World Bank has expanded its environmental staff from a total of five to more than 200 (*Economist*, 25 Dec 1993), and its level of purely environmental lending was $1.6 billion in 1991, by comparison with $400 million in all previous years (P. Birnie, 'The UN and the Environment', in Roberts and Kingsbury, eds, *United Nations, Divided World*, 2nd edn, Oxford University Press, Oxford 1993).

the conference 233

highly significant development that will be examined in more detail in the next chapter. Finally, it has given UNEP a boost in UN standing (and, it is to be hoped, finances) some element of which again may be lasting.

Even at the level of its written conclusions we have seen that Stockholm should not be completely written off. It is striking how much of the environmental business of the subsequent 20 years – from CITES to climate change – is foreshadowed in those conclusions, and the famous principle 21 has hovered at the edge of international law throughout the period. It is not unreasonable to hope that the same fate may await some of the UNCED output. Certainly Rio's adoption of the precautionary principle, the polluter pays principle and Stockholm's 'non-principle 20' will give those texts some new force in international environmental business in years to come. More generally, Rio's espousal of the principle of sustainable development at least suggests the direction in which policies that reconcile the environmental concerns of the developed world with the determination of developing countries to maintain their rates of economic growth may be sought.

> **One other lesson taught by looking at the Rio and Stockholm Conferences together ... is that the international system has a very short memory ... one has the impression of the Rio negotiators (and observers) laboriously rediscovering the roadblocks that made Stockholm such hard work .. it is difficult to resist the feeling that progress would have been greater if the participants had known a bit more of the history.**

One other lesson taught by looking at the Rio and Stockholm Conferences together (and which has been implicit in a great deal of the above) is that the international system has a very short memory. Here were two global conferences, a mere 20 years apart, covering very similar subject matter, organized under the same international auspices and even with the same Secretary General. In a bureaucratically and politically perfect world Rio would have picked up where Stockholm left off and carried the work some way further. Indeed, there were elements of continuity, most of them provided by Maurice Strong himself, such as the use of the Agenda 21/Earth Charter formula to build on the Stockholm Action Plan/Principles. But in general one has the impression of the Rio negotiators (and observers) laboriously rediscovering the roadblocks that made Stockholm such hard work. The Stockholm precedents and conclusions (with the inevitable exception of principle 21) were strikingly little quoted during the preparation for Rio and there are eerie parallels between the difficulties the two conferences erected for themselves, from the rediscovery of the central North-South differences on population, forests and financial assistance to the 'Dresden china' delicacy of treatment required for the Stockholm Principles/Earth Charter. Of course there was some progress, both intellectual and substantive, between the two conferences – perhaps best illustrated by the adoption of the concept of sustainable development as a way round the central confrontation of Stockholm between economic growth and environmental protection – but it is difficult to resist the feeling that progress would have been greater if the participants had known a bit more of the history.[11]

> **The brutal fact is that the US completely mishandled its approach to Rio and the linked negotiations and earned worldwide brickbats as a result ... it will be important for progress on the world's lengthening environmental agenda that its richest and most influential country finds its way to a less adversarial approach than it managed at Rio.**

The very different performances of the US and the European Community in the Rio process merit some comment. The brutal fact is that the US completely mishandled its approach to Rio and the linked negotiations (and with the change of administration, it has since abandoned some of the most controversial positions it took at the time) and earned worldwide brickbats as a result. The member states of the EC, on the other hand, many of which had problems similar to those identified by the Americans on such central issues as increased aid for the developing world, the deficiencies of the biodiversity convention, and high-cost action to tackle climate change, nevertheless man-

---

[11] Indeed, the author's consciousness of his ignorance in that respect was one of the motivations for writing this book.

aged to emerge from the process with their bridges to the rest of the world reinforced rather than undermined, and their influence and standing enhanced rather than eroded. There are a number of explanations for this outcome, many very local and conjunctural, such as the right-wing challenge to President Bush in the Republican conventions which were under way at the same time. There was, however, also a larger factor at work which has implications for international environmental cooperation in the future. The EC nations, by virtue of their size and proximity, are by now well adjusted and attuned to doing environmental business by international negotiation. They have learned to be attentive to international currents of opinion and to be ready to look for the compromises necessary to get an agreed result. The US has not yet learnt this lesson to anything like the same degree. Its environmental policy-making is much more exclusively focused on domestic politics than is true for any state in Europe. This has led on occasion to a strikingly introverted approach to external environmental problems (as also illustrated by the acid rain debate with Canada and arguments with Mexico over dolphin protection). It is, of course, unsurprising to see large countries using their weight to protect their interests. It is, however, rather more surprising to see them do this in a way that, by any reasonable measure, actually damages their (and perhaps the world's) interests, as happened at Rio. It will be important for progress on the world's lengthening environmental agenda that its richest and most influential country find its way to a less adversarial approach than it managed at Rio.

The final point to make about Rio is simply its size. It is hard to imagine a more convincing demonstration of the distance environmental issues have come in Western, and increasingly world, politics. Since their birth in the 1960s they have moved from the margins to their present position as the subject matter for the first ever world summit. Quite what this has to say about our ability to tackle them in the future we will consider in the next chapter.

Chapter 11

The Future

> *We are as Gods, and might as well get good at it.*
> THE WHOLE EARTH CATALOG

11.1 The Challenge

The question raised at the start of this book was: 'is the international system as at present constituted capable of responding to the threat of global environmental degradation?' The history of the past 20 years can be used to offer a gloomy answer. Over the period, as we have seen, world population has increased by over 40%, and human pressure on natural resources has risen significantly faster than that (oil consumption, for example, is up by 60%). As a result human activities now constitute a significant, and often destructive, component of the earth's ecosystem. Some sources suggest that forty per cent of the entire natural photosynthetic product of the biosphere is appropriated by man with, as one side product, other species being rendered extinct at thousands of times the natural rate. Twenty per cent of the carbon dioxide in the atmosphere has been put there by man, threatening profound changes in the global climate. Over the past 50 years the productivity of more than 1.2 million hectares of land, an area larger than China and India together, has been significantly lowered by human agricultural and other activities. At current rates of destruction the world's remaining tropical forest may be gone in 50 years.

It is against this background that a number of commentators have argued that the existing international system of sovereign states is simply inadequate to tackle the environmental problem. Back in 1976 Arnold Toynbee wrote: 'The present day global set of local sovereign states is not capable of... saving the biosphere from man made pollution or of conserving the biosphere's nonreplaceable natural resources... what has been needed is a body politic on a scale in which the participants... would be citizens of the world state.'[1] This sentiment has echoed through environmental literature up to the opening words of the Brundtland Report: 'The Earth is one but the world is not. We all depend on our biosphere for sustaining our lives. Yet each community, each country, strives for survival and prosperity with little regard for its impact on others.'[2]

The logic, which as we have noted dates back to Machiavelli, is that states will pursue their own individual national interests and thus find it extremely difficult to cooperate for the common good. Moreover, the obstacles to such cooperation grow higher as the costs and the number of states that are involved rise. The common good becomes more remote from the interests of any individual party, the capacity of the system to monitor and enforce obligations on each individual state grows less, and the attraction (and possibility) for each state of 'free-riding' on action by others grows greater. There are in addition, as we have amply seen in the climate change and biodiversity cases, the sheer *mechanical* problems of negotiating a worthwhile agreement among 180 countries. Thus, it is argued, the prospects of effective joint action by all 180 states of the world to tackle the global problems which are now emerging are vanishingly low.

Discouraging though this line of argument is, there is little point in basing our hopes for the future of the planet on the disappearance, or conversion to global altruism, of the nation state. Even with fast-rising levels of global interdependence – exemplified for example by booms in international communication, cultural exchange, trade and travel – the nation state evidently remains the primary locus for the political loyalties of the vast majority of the people of the world. It is not rapidly going to vanish. That most sophisticated experiment in edging it towards disappearance at a regional level, the Euro-

---

[1] A. Toynbee, *Mankind and Mother Earth: A Narrative History of the World*, Oxford University Press, Oxford, 1976.
[2] For a list of other references in the same sense see P. Haas, R. Keohane and M. Levy, 'The Effectiveness of International Environmental Institutions', in Haas, Keohane and Levy, *Institutions for the Earth*.

pean Community, has regularly been set back by resurgent particularism in individual member states; and even if it went as far as its most enthusiastic proponents could wish, would reduce the number of nation states in the world by less than 5%. Nor is there much sign that environmental pressures (even at their most extreme, as in Bangladesh or Ethiopia) are causing the nation state to disappear.

Moreover, as long as environmental agreements are negotiated by nation states it seems certain that national interest will play a large part in deciding the outcome. Examples abound in the preceding pages, from Brazil's exclusion of Principle 20 from the conclusions of Stockholm through French resistance to any mention of radioactive pollution in the Mediterranean agreements to Malaysia's evisceration of forest controls, and America's evisceration of carbon dioxide emission controls, at Rio. A particular and enduring instance, highly relevant to the future prospects of global environmental collaboration, has been the resistance of the developed countries – at Stockholm, through the NIEO, and at Rio – to demands that they increase their financial and technological transfers to the developing world over and above what they see as being necessary to tackle environmental problems directly relevant to themselves. New instruments have been created – UNEP's Environment Fund, the ozone layer fund, the Brazilian forest project, the GEF – but only in response to pressing political concern in developed countries about particular issues. Efforts to achieve a general increase in aid, or substantial funding for issues principally of concern to developing countries (such as desertification) have regularly fallen flat. There is no reason to believe that this situation will change in the near future.

It is therefore clear that the retreat from nationalism and anthropocentrism which 'dark green' environmentalists have insisted is necessary is not an early prospect. Even in the West, green parties have been more the bearers of a minority creed than serious contenders for national political power, and in the South of course popular environmentalism has been weaker still. Economic growth will for the foreseeable future remain a (if not the) central objective for the countries of the world.

Happily, a closer look at the environmental problems associated with this growth (such as those discussed in the preceding chapters) does not suggest that we are faced with imminent catastrophe. Early claims that we are rapidly approaching unbreachable 'limits to growth' as we deplete the earth's stock of non-renewable resources have proved wrong. Since the Club of Rome débâcle it no longer seems plausible, at least over the time frame where prediction makes any sense at all, to argue that growth will be hindered by global natural resource shortages. Any such shortages as appear are likely to be accommodated, as in the past, by price changes, substitute materials and new technology.[3] This does not mean that there are no limits to growth – we certainly live on a finite planet – only that, on the evidence currently available to us, those limits remain so remote as not to be a sensible concern for public policy.

This leaves the problems of pollution and ecosystem destruction upon which environmental concern has increasingly focused since the Stockholm Conference. Inevitably, as economic growth proceeds, these problems are going to require increasing attention and action. Already, as the scale of human economic activities has more closely approached the scale of the natural cycles of the planet, we have seen the size of the resulting environmental disruption rise from the local (urban smog and polluted water courses), through the regional (acid rain and the quality of the Mediterranean) to the truly global (the ozone layer and climate change). We have however, also seen, rather contradicting the view that a disunited world is doomed to a 'tragedy of the commons', the emergence of national and international responses to these

---

[3] This is not a universally shared view. There has been a continuing stream of 'limits to growth' literature. In a fairly characteristic example, H. Daly (*Steady State Economics*, 2nd edn, Earthscan, London 1992) asserts the 'theorem' that it is impossible for the whole world ever to achieve a Western standard of living (because e.g. the US with 6% of the world's population accounts for one-third of the world's consumption of non-renewable resources). This approach, however, seems to overlook the implications of technological progress. If, for example, over the next hundred years the world's population followed the UN's middle projection and stabilized at double its present level, living standards worldwide rose to current Western levels, and at the same time the best Western levels of energy efficiency were extended worldwide, and if those standards improved at a (by recent standards modest) rate of 1% per annum, then global energy consumption in the year 2094 would be just 10% higher than it is now. Of course, this uncomplicated calculation deserves no more respect than any of the other predictions currently on offer (Foley, *The Energy Question*, Ch. 14 is devastating on the persistent massive erroneousness of long-term forecasts of energy demand), but it does raise doubts about any 'impossibility theorem' asserting the predestinate insufficiency of the world's resource base to provide high standards of living for all.

The future

threats. And some of those responses are already at level three. What factors have contributed to this?

11.2 The Benefits of Prosperity and Democracy

First and foremost must come the benign combined effects of prosperity and democracy. Once a people has achieved a certain level of material affluence it turns its attention from meeting immediate needs to the longer term – and the longer term, almost by definition, includes the environment.[4] Instead of pressing hungrily for the next crop, farmers can begin to think about topsoil loss. A tree becomes a guarantor of next year's local rainfall pattern instead of tomorrow's firewood. Sulphur dioxide becomes a real cost in terms of acid rain rather than a negligible byproduct of much-needed electrification. Moreover as countries become richer they acquire the administrative systems to enforce environmental controls and thus protect the common good against individual self-interest.[5] Democracy, as we have already noted, is crucial in transforming popular concern about the environment into political action. Even with the 'greening' of some industrial attitudes we have continued to see special interest groups lobbying against environmental legislation which they view as disadvantageous to them. Democracy creates a counterweight to such producer pressure. The influence of these factors has been a consistent strand of our story. The absence of prosperity has produced developing country resistance to agreements which they see as constraining their prospects for economic growth, ranging from marine pollution to climate change. Weak administrative infrastructure has also inhibited their ability to enforce the

---

[4] There is a large economics literature on this lengthening of the time horizon with growing affluence. For a summary with further references see D. Pearce and R. Turner, *Economics of Natural Resources and the Environment*, John Hopkins, University Press, Baltimore 1990.

[5] This does not mean that rising prosperity of itself solves all environmental problems. We have already seen (in Chapter 4) that while rising income levels are correlated with falls (sometimes after an initial rise) in the levels of many pollutants, they have also been accompanied, so far, by steady rises in others (e.g. waste levels; cf. World Bank, *World Development Report 1992*). What does seem clear, however, is that rising prosperity has been accompanied by growing political and public consciousness of environmental problems, and rising economic and administrative capacity to deal with them.

environmental laws they have introduced – as we saw in particular in the cases of India, Brazil and Mexico. The absence of democracy had its worst environmental consequences, which it will take decades to remedy, in the Soviet Union and Eastern Europe. But given prosperity and democracy we have seen popular lobbies take on, and beat, industrial resistance on issues as various as the North Sea and the ozone layer.

The effect of these various forces has been at its most apparent in the developed West. The past 20 years have seen the environment take a permanent place in Western political and economic discussion. Environmental concern is now an uncontentious component of the daily business of government, the media, education, industry and hundreds of millions of individuals. Of course, the prominence of the subject will vary, as demonstrated by the two booms of popular environmentalism in the late 1960s and the late 1980s. But behind such temporary fluctuations it seems clear that there has been a permanent shift by Western society into a much greater sensitivity to, and determination to combat, environmental deterioration. This shift has had a striking effect on public debate in the West. In such debate those who espouse causes in which few believe risk being branded eccentric or extremist; while those who argue firmly for the conventional wisdom are viewed as sound and dependable figures. The effect of the shift in the public standing of the environment since the 1960s has been to transform it from the first type of issue to the second. It is now those who argue against environmental concern who are the eccentrics.

Thus the period has seen these countries build, from a very low base, a substantial edifice of environmental action and collaboration. The vast majority of this action has been, as befits the electorally driven nature of the subject, to tackle domestic pollution. Indeed, strong evidence of the difference in priority given to domestic and international problems in the case of the US is provided by that country's willingness, in 1990, to adopt Clean Air Act amendments which would cost an estimated $25 billion per annum to implement, while at the same time resisting internationally the establishment of a fund crucial to the protection of the ozone layer whose cost to the US was less than 0.4% of that. The major European countries have also led with domestic action, although as they are smaller, many more problems from their point of view are international in scope and require international collaboration (as exemplified in particular by the European Community's huge output of environmental legislation). Japan too started with domestic issues (although the popular pressures which these aroused rapidly subsided and Japanese environmental activism now seems to be led by a growing 'green' industrial lobby keen to export).

Indeed, the various pressures on developed country governments – press, popular opinion, scientific views, international views – may now be such, on certain topics, as to cause them to overproduce environmental regulation. In the US the $20 billion programme to remove asbestos from public buildings has been officially admitted to be a mistake and the 'superfund' programme has been extensively criticized as a grotesquely exaggerated and expensive attack on the problem of hazardous waste dumps.[6] Federal government estimates show that certain drinking water standards cost $92 billion per premature death averted.[7] EC drinking water controls have been attacked on similar grounds, as well as for the sheer unattainability of some of the standards (e.g. for pesticides) set in them. At the international level we have seen that some of the controls agreed on North Sea dumping have a very flimsy scientific base. Popular concern at the time of Chernobyl produced EC radioactivity contamination standards for foodstuffs much tougher than scientists recommended. While there is now widespread agreement among scientists that world stocks of the minke whale are large enough to allow some whaling, there is very little sign that most Western governments are willing, in the teeth of popular opposition, to accept this view and vote in the International Whaling Commission to end the moratorium currently in force.

On the general issue of whether political pressures are moving ahead of scientific knowledge, the EPA's Science Advisory Board concluded in 1990 that environmental laws 'are more reflective of public perceptions of risk than of scientific understanding of risk'. The then head of the EPA, William Reilly, commented: 'We need to develop a new system for taking action on the environment that isn't based on responding to the nightly news. What we have had in the US is environmental agenda setting by episodic panic.'[8] There are arguments on the other side. Some actions taken ahead of scientific con-

---

[6] *New York Times*, 21 March 1993.
[7] 'Office of Management and Budget; Cost Benefit Anaysis of Certain Environment and Safety Regulations', US Government, Washington, DC 1991.
[8] *New York Times*, 21 March 1993.

firmation have been abundantly vindicated subsequently; the CFC phaseout and European sulphur dioxide emission cuts are two instances. The small initial action may be a vital step in reorienting industrial attitudes to permit larger actions later, as in the case of ozone and (arguably) climate. But overall it is difficult not to conclude that Western public concern about the environment (and relative indifference to the advice of economists) is producing standards which in many cases are tougher than is justified by the problem.

There remain of course many areas – use of agricultural fertilizers and control of waste immediately leap to mind – where a great deal *more* action is still needed. But the clear implication is that, as regards the environment within developed countries, and in regions dominated by developed countries (such as Western Europe and the North Sea), the problems are either already on the way to solution or are arousing enough political concern to make progress probable. As economic growth proceeds, new problems will emerge. But, in the developed West, threats to the environment now look more like an endemic and recurrent complaint of modernity – like inflation, or urban blight – than a fatal disease; a regular stomach upset (requiring, of course, appropriate treatment) rather than intestinal cancer.[9]

11.3 Developing Countries

The situation in most developing countries is radically different from that in the West. Popular environmental concern, where it has existed at all, has been confined to local problems. The environment has usually been a government-driven 'top-down' issue, with governments giving a high priority to economic development and the environment taking very much second place. It is politically inconceivable that developing country governments will for environmental reasons abandon their efforts to achieve higher standards of living, and all of the environmental negotiations we have looked at have underlined this fact.

---

[9] A judgement which is considerably reinforced by the current 'dematerialization' of Western economies. For example, the energy and raw material content of a unit of Japanese production fell by 40% between 1979 and 1984 (MacNeill, Winsemius and Yakushiji, *Beyond Interdependence*). This trend, which is apparent throughout the West, plainly bodes well for levels of pollution, waste generation and raw materials consumption in developed countries.

The future

We have nevertheless seen some evidence in the developing countries too of the 'international contagiousness' of environmental action which has been so apparent in the developed countries. The speed with which, following Stockholm, developing countries such as China and India set up environmental ministries and embarked on the business of environmental lawmaking is clear evidence that the intellectual, scientific, media and official networks which have enabled the West to move forward broadly in unison on environmental issues also operate, albeit more weakly, between North and South. At the international level this has facilitated developing country cooperation in, for example, CITES and the Mediterranean agreements.

It is thus possible to take a less pessimistic view of the environmental prospects for the developing world (with one conspicuous exception) than that suggested at the start of this chapter. As the developing countries becomes richer and more democratic (as, happily, most of them are; average GDP per capita in the developing world rose by 3% per annum from 1965 to 1989, and the fall of communism has been followed by a 'wave of democratization' in the third world[10]), so, as in the case of the West, the attention they devote to their environmental problems will grow. Indeed, as we have seen, this process is already well under way in such industrializing countries as Greece, South Korea, Israel and Singapore. Moreover even countries at a lower level of prosperity, such as India and China, have been able to draw on the increased knowledge produced by the past 20 years of global environmentalism to tackle the environmental roots of such problems as land degradation and water borne diseases, so that the South has seen a steady expansion of, for example, population control and water sanitation programmes.

In other words the growth of environmental knowledge and awareness of the past 20 years, even if principally concentrated in the developed countries has helped developing countries to identify environmental problems which it is in their own interests to tackle, and they are accordingly (subject always to their administrative and financial means) setting about doing this. It follows that at the domestic level that the large proportion of the developing world which is growing more prosperous can be expected gradually to bring more attention and resources to bear on environmental protection. Ultimately they

---

[10] *World Development Report 1992*; S. Huntingdon, *The Third Wave*, University of Oklahoma Press, Oklahoma 1992.

can be expected to turn around the environmental degradation which many of them are experiencing – that is to say they will follow trajectory similar to that followed by the West, with the early phases of growth being accompanied by significant environmental damage, which then begins to be tackled as growth takes off. There is reason to hope, too, that since they can draw on the experience (and to some extent the resources) of the West, together with the 'sustainable development' lessons articulated by Mrs Brundtland, the environmental nadir of their growth curve will be considerably less polluted than that of Western environmental history.

One factor which may well help substantially here will be the influence of Western investment. This point has received surprisingly little attention considering that the levels of official aid so fervently argued about at Rio are now significantly less than Western private investment in the third world. Since, as we have seen, Western shareholders and companies increasingly expect their projects in developing countries to meet environmental criteria comparable to those prevailing at home (and are conscious of the eagle eye of the environmental NGOs if they slip too far from that standard) this flow of funds can be expected to carry higher third world environmental standards with it.

The black spot which needs to be added to this relatively sanguine picture of developing country prospects is the situation of the least developed countries, notably in south Asia and sub-Saharan Africa. Prosperity in these countries is often *not* growing, so the mechanisms identified above under which rising developing country prosperity raises both the will and the capacity to tackle environmental degradation cannot operate here. Indeed, such countries are often being dragged down by the very ecological degradation they do not have the resources to solve. A conspicuous example is the lethal combination of rapid population growth, overintensive agriculture and consequent soil degradation in some African countries. In countries like Zaire, Uganda, Niger and Zambia GDP per capita has fallen by an average of more than 2% per annum over the past 25 years.

These are not problems in which the West has any evident self-interest (other than moral) in helping to tackle as their impact is purely local. We have, moreover, already noted the historic unresponsiveness of the West to generalized calls for increases in aid. Nor will the rise in Western private investment

The future

in the third world help much. Most of this is going to the richer developing countries. There are palliatives: television pictures of famine often generate some temporary assistance, and, as we have seen, the NGOs now have quite large funds which can often be used to help. But the core issues of population growth and agricultural impoverishment which hang catastrophically over the heads of such nations as those mentioned above are not of the sort which can easily be solved by a system of nation states acting exclusively on the basis of individual self-interest.

11.4 The Global Problems

The fact that both developed and most developing countries can ultimately be expected to deal with their own domestic problems (and, by extension, regional ones too – a direction in which the embryonic environmental efforts of such bodies as ASEAN, the Amazon Pact and the Southern regional seas agreements are already pointing) still leaves a question mark over the big global problems. But on these, too, our history suggests that it would be a mistake to be too pessimistic. The prospects are not so much black as extremely clouded. The entire international community *has* succeeded in cooperating to tackle certain global problems – CITES and the ozone layer agreements are examples, and the Mediterranean agreements are a compelling display of North/South cooperation to tackle a shared environmental problem on a smaller scale.

Moreover, as we have seen, there has been a degree of progress, in a remarkably short time, on the issues which now dominate the agenda – deforestation, biodiversity and climate change. Despite the intense scientific work which has gone into them, and continues to go into them, these issues remain at a relatively early stage in their definition. We have seen that even from the developed countries' point of view it is not at all clear how much CO_2 emissions limitation is currently justified.[11] From the point of view of the developing countries the situation is, if anything, less clear. On the one hand they are confronted by a potentially enormous, but almost entirely unquantified, threat to their large agricultural sectors and to the viability of their natural ecosys-

---

[11] Although the precautionary principle plainly argues in favour of some limitation.

tems; on the other they face calls for action on, for example, fossil fuel burning which could seriously inhibit their prospects for economic growth. On deforestation and species destruction, the point has legitimately been made that the past 20 years of steady depletion of the tropical forests and the species they contain has been accompanied by steadily rising global, and in most cases local, prosperity. The proof is not yet in that now is the time to stop. The fact that, in these circumstances, agreements (however thin) were achieved at Rio on these issues was therefore an impressive demonstration of the global political concern they have already aroused.

Biodiversity and forests

The avenues for progress on climate and biodiversity (which in effect includes deforestation) seem likely to diverge in the future. On biodiversity and forests the Rio negotiations heavily underlined the limitations of the multilateral approach. The fact is (as a group of key developing countries have consistently emphasized) that we are talking here about conservation of resources within natural borders. The most effective and flexible way to do this is much more likely to be through individual arrangements with the countries concerned than through global agreements. This does not preclude some overall global strategy, nor the development of multilateral funding instruments, such as the GEF. But, as the history of the World Conservation Strategy underlines, such strategies and instruments, while perhaps creating some political impetus, are finally only as good as the individual projects to which they give rise. There are indeed signs, of which the Brazilian forest project and the reform of the TFAP are the most conspicuous, that this 'renationalization' of action on forests and biodiversity is already underway. Nor is such renationalization confined to official action. The Costa Rican Government has recently established an innovative arrangement with the Merck pharmaceutical company to try to turn its biodiversity to profit,[12] and if the world's untapped reserves of biodiversity are genuinely as potentially valuable as

---

[12] For more details, see W. Reid, S. Laird, R. Gomez, A. Sittenfeld, D. Jonzen, M. Gollin and C. Juma eds, *Biodiversity Prospecting: Using Genetic Resources for Sustainable Development*, World Resources Institute, Washington DC, 1993.

their advocates claim one would expect to see entrepreneurs coming forward with other similar schemes to try to share in the benefits.

The big question hanging over the future of the forests and biodiversity is thus not: 'How fast will the big international conventions progress?' It is a combination of much more local questions. How fast will the present rate of destruction be moderated by recognition in individual developing countries of their own domestic interest in maintaining some forest cover and species abundance (as expressed in the growth of deforestation legislation and conservation programmes)? How fast will it be moderated by rises in Western political and economic concern (as expressed through more public and private funding for biodiversity and forest conservation)? It is of course impossible to tell, but there have recently been rapid moves in the right direction. Brazil and Indonesia now have more protected areas each than there were in the entire world 20 years ago. With rising world food production the economic benefits of clearing forests and other natural habitats for agricultural purposes are diminishing.[13] It is becoming clear that developing countries are undercharging for logging licences, so licence fees (and the incentive to reafforest) are likely to go up and the rate of logging down.[14] And there are signs, too, that as rural population growth diminishes so does deforestation.[15] Developing countries are also increasingly identifying policy options which are both ecologically and economically beneficial, such as Brazil's abolition of subsidies for deforestation and the growth of ecotourism. None of this means that the problem is going to disappear overnight. The world is going to lose a great deal more tropical forest and many more species before the tide turns (and it was in any case always unrealistic to expect that, say, Brazil would keep two-thirds of its land area as virgin forest at the behest of developed country populations that had grown rich by cutting their forests down). It is, however, strongly arguable that the factors are now falling into place that will bring about that turn of the tide.

---

[13] T. Swanson, 'Conserving Biological Diversity', in D. Pearce, ed., *BluePrint 2: Greening the World Economy*, Earthscan, London 1991.

[14] Paul Harrison, *The Third Revolution*, Penguin, London 1992, notes that such undercharging cost Indonesia $3 billion between 1979 and 1982.

[15] On this basis, P. Harrison (ibid.) estimates that deforestation should stop in Asia early in the next century, and continues now in South America only because of subsidies.

Climate change

On the face of it the climate change issue is much harder. This is of course an issue of protecting the global commons rather than individual national resources. Thus independent action through individual national self interest will not be enough. The problem has to be tackled, as in the cases of the oceans, acid rain and ozone, through multilateral agreements. The basis is now there in the form of the climate convention. This is likely to give rise to rather more concrete action than the polemics of the negotiation foreshadowed, not least because (as in the case of biodiversity) there are a number of 'no regrets' policy actions that developed and developing countries could take which will help on both climate change and economic growth grounds. Indeed, precisely because of all the argument about targets and timetables, the very substantial scope for policies of this type has been rather neglected. If, for example, the entire OECD (and in particular the US and Canada) could raise their energy efficiency to Japanese levels, OECD energy consumption would fall by over a third, and carbon dioxide emissions proportionately. The scope for improved efficiency in the developing world is even higher. Energy consumption per capita in these countries will continue to rise with industrialization; but if energy consumption per unit of GNP could be brought down even to the relatively unambitious European or US levels this would constitute a 50% improvement in energy efficiency in India, and an up to 80% improvement in China.[16]

However, given the level of scientific uncertainty, and therefore of uncertainty in each nation state about its own interests, progress beyond such 'no regrets' policies is likely to be slow unless either some event on the scale of the discovery of the 'ozone hole' makes the dangers much clearer to everyone concerned or at least we attain much greater scientific clarity on the potential costs (and benefits) of climate change to various areas of the world.[17] Such

---

[16] Current expectations are in any case for improving energy efficiency in both developed and developing worlds (IEA, *World Energy Outlook to 2010*; OECD Paris 1993). These possibilities also take no account of the potential for sharply improved energy efficiency technologies (see J. Leggett, ed., *Global Warming: The Greenpeace Report*, Oxford University Press, Oxford 1990).

[17] See also E. Skolnikoff: 'The Policy Gridlock on Global Warming', *Foreign Policy*, Summer 1990, for other arguments pointing in the same direction.

clarity is unlikely to be available quickly. Since climate is an issue whose effects are likely to be felt only over decades, if not centuries, the delay this will probably cause for international action, while having real costs (e.g. by forcing species to adapt faster, and so probably raising rates of extinction) should not be disastrous.[18]

Even when the information does become available, and even if it does unequivocally confirm the need to cut greenhouse gas emissions, the prospects remain uncertain. One can confidently predict that the costs of effective action will be a great deal lower than some of the enormous numbers currently being thrown about (mostly by those opposed to action). Over and above the substantial prospects for 'no regrets' actions, the experience of high-cost environmental legislation in the past

> **The whole history of international environmental action has been of arriving at destinations which looked impossibly distant at the moment of departure.**

has been that economic agents tend to find much less expensive means of meeting its requirements than they originally predict.[19] Even so, achievement of the convention goal of stabilizing carbon dioxide concentrations in the atmosphere, if it turns out that that is what is really required, will almost certainly demand the largest international agreement of any sort ever negotiated. It would probably have to combine substantial commitments by all countries on the future path of their greenhouse gas emissions (with all of the implications that carries for future economic growth and structure) with un-

---

[18] e.g. W. Nordhaus ('How Much Should We Invest In Preserving Our Current Climate?', Yale preprint 1992) estimates, on the basis of admittedly very incomplete data, the cost of a ten-year delay in action as 0.003% of world consumption. M.E. Schlesinger and X. Jiang ('Revised Projection of Future Greenhouse Warming', *Nature*, 350, March 1991) reach a similar conclusion. This work does not take into account the possibility of climatic 'shocks' -i.e. changes of climate more abrupt than would be expected from the steady build-up of greenhouse gases in the atmosphere; but we currently know very little about the likelihood or possible magnitude of such shocks, so even the precautionary principle tells us little about how much insurance it would be right for mankind to take out against them.

[19] This has, for example, been the consistent experience of automobile and power-generation companies in meeting antipollution requirements.

precedented Western commitments on aid and technology transfer. All that can be said (on the negative side) is that the scale of the thing at present levels of international cooperation looks daunting. Certainly it will not be achieved all at once, but rather through the evolutionary sequence of manageable steps we have already seen in the case of the ozone layer and ocean pollution. On the positive side, however, the whole history of international environmental action has been of arriving at destinations (such as the acid rain and ozone layer agreements) which looked impossibly distant at the moment of departure.

11.5 Forces for international cohesion

The success of environmental negotiators in reaching such apparently impossible destinations lies in a number of factors conducive to international environmental cohesion which have evolved over the past two decades, and which help give current environmental business its rather special flavour.

Negotiating processes

The first of these factors has been the combination of the use of environmental negotiating *processes* (as opposed to definitive agreements) with what might be called the internationalization of environmental public opinion. On such issues as the North Sea, European acid rain and the ozone layer we saw that the initial agreement amounted to no more than a 'toe in the door' whose substantive content was low but whose principal effect was to give rise to a succession of meetings which were used as an opportunity by those countries with environmentally more advanced positions to place public pressure on the others, both directly and via environmental NGOs, to shift in their direction. In all the cases cited, together with many European Community negotiations, this process, which is evidently dependent upon arousing environmental public opinion in the countries upon which pressure is being placed, had the effect of moving backmarker governments into the mainstream. This simply would not have been possible without, first, the globalization of news which has meant that the ozone layer can be a headline simultaneously in 20 countries, and second, the uniquely open and public style of international environmental negotiation, which was one of the key innovations of Stockholm and which means that NGOs and others can apply their pressure on the right issues at the right time. This process has depended crucially on some states (such as those of the 30% Club) being ready to go ahead *even though they don't know others will follow* – a point to which we will return below. Finally, an important component of the success of this 'process' style of negotiation has been the opportunity it has given for the science to progress and for industrial attitudes to adjust as the regulatory environment evolves. The ozone layer negotiation was a particularly compelling example of this. At a key 'breakthrough' point (the Montreal Meeting) the science hardened up significantly and industry accepted that it was going to have to adjust to a new regulatory framework; it thereupon became a force for progress rather than delay.

The downside of this stately 'toe in the door' procedure of assembling acts of international collaboration to deal with particular environmental issues is that it takes time. In the case of European acid rain it was 15 years from the first scientific concerns to the sulphur protocol, in the case of the ozone layer it took ten years of international activity to produce the Montreal Protocol. Over the period while the science is firmed up and the political will assembled to take effective action on a particular problem the damage will often continue to grow. There can be real environmental costs associated with such delay. We are going to lose a great deal more forest and biodiversity before the situation stabilizes. It is estimated that with the number of CFCs already in circulation the ozone hole will not begin to mend until the middle of the next century. One can even conceive of circumstances (such as carbon dioxide levels rising to the point where they touch off a major instability in the earth's climatic system, such as triggering another ice age) where delay could produce irreversible environmental degradation. Such cases argue firmly in favour of the 'precautionary principle'. But in general the experience has been (whatever the assertions of urgency at the time of the negotiation) that no environmental damage is irreversible,[20] and once a repair effort is energetically launched the damage to the ecosystem can often be undone with surprising rapidity – as has been the case for example with the London smogs, the Great Lakes and the survival of the East African elephant, and is begin-

---

[20] Although massive extinction of living species may be an exception to this.

ning to happen with Mediterranean pollution levels and European acid rain. Claims of urgency (often pressed by NGOs) have an evident political value in creating an atmosphere where governments find themselves under pressure to concede. This has been an effective tactic in the acid rain, ozone layer and North Sea negotiations. But such claims are often scientifically weak and on occasions have been transparently ridiculous, as with the criticism of the UK for choosing an original target of 2005, instead of 2000, for its stabilization of carbon dioxide emissions to deal with a problem whose effects were anticipated for perhaps a hundred years hence.

Environmental science
The second factor for cohesion has been international science. There is a paradox here. Environmental problems are about as far as it is possible to get from the simplicity of the laboratory bench. They tend to be highly complex and open-ended, require inputs from a number of different scientific disciplines and lend themselves, at best, to highly uncertain assessments of their causes and predictions of their future development. In ordinary circumstances such material does not normally lead to vigorous public action. The very real uncertainties and the occasional maverick opinion give those governments which for other reasons may want to go slow the best possible justification for doing so – as we have seen the UK do in the case of acid rain and the US in the case of climate change.

It is therefore striking, and amounts to something of a revolution in the sociology of science, that the world's scientists, aided and abetted by such organizations as UNEP and the WMO, have increasingly taken to coming together (in what can only be described as a highly political manner) to produce agreed assessments of environmental problems. Scientific figures, such as the chairman of IPCC Working Group I, have emerged who can put the conclusions over, despite all the uncertainties, in language clear and forceful enough to elicit a policy response. This process now has sufficient policy impact to have given rise to occasional efforts to bend scientific advice, as was seen in the European Community during the argument about acceptable radiation levels in foodstuffs following the Chernobyl explosion, and in the amendment by the White House of Hansen's Congressional testimony on climate change.[21]

Indeed the power of a united scientific view to push even unwilling governments into action is now one of the key mechanisms of international environmental cooperation. We have seen that in the cases of ozone, the Mediterranean and climate change the construction of an international scientific consensus on the nature of the problem was the first step in manufacturing a political consensus on what should be done about it. Governments, naturally, tend to believe their own scientists more than others, so the international scope of the consensus, and in particular its inclusion of scientists from the central countries in the negotiation, can be vital. Some of the countries in the Mediterranean negotiation held back until their own scientists were convinced of the problems,[22] and it was precisely to ensure coverage of this kind that the IPCC went to such efforts to draw in developing country scientists.

This growing influence of scientists in environmental policy processes (which on the whole have tended to be dominated by lawyers and economists[23]) should help produce better national and international environmental decisions. But there are points to watch. The first lies in the possibility of corporate scientific error. It has been rightly pointed out, for example, that in the case of climate change the community of qualified scientists is relatively small and incestuous, and that an 'inverted pyramid' of potentially massive implications and costs rests upon the views of these few experts.[24] It could be very expensive for all of us if they jointly went astray.

The second problem lies in the link between science and public views. Democratically elected governments may listen to their scientists, but they are elected by their publics. Public understanding of the science of any particular environmental issue is inevitably thin and is almost certainly weighted in the direction of the most sensational (and so most publicized) predictions, as illustrated by the impact of Jim Hansen's climate change testimony to the US

---

[21] This is not an entirely new phenomenon; cf. Ibsen's 1882 play, *An Enemy of the People*.
[22] Haas, *Saving the Mediterranean*, Columbia Uiversity Press, New York 1990. Indeed it was Haas who came up with the idea of 'epistemic communities' of scientists imposing convergent pressures upon governments.
[23] L. Caldwell, *Between Two Worlds*, Cambridge University Press, Cambridge 1994.
[24] E. Skolnikoff, 'The Policy Gridlock on Global Warming'.

Congress. To the extent that this public jumpiness gives political potency to the precautionary principle (and indeed helped get governments to focus on environmental issues in the first place) it is beneficial, but we have seen a number of cases above, eg on nuclear matters, where it seems now to be producing environmental overregulation. This problem is being compounded by the growing technical sophistication of science itself, which is able to identify pollutants in steadily more minute quantities.[25] Sometimes the resulting public alarm is justified. The technical developments which led to the discovery of the ozone hole galvanized the world into action on a very real environmental problem. Sometimes, however, it is not. For example, the pesticide limits in the European Community's drinking water directive were arbitrarily set at the (then) limit of measurability, and have since been found to be both unachievable and unjustified. The growing precision of science is thus double edged: it reveals environmental threats of which we were not previously aware, but can also create the impression of new environmental threats where there are none.

The NGOs

The third major factor contributing to international convergence on environmental issues is the environmental non-governmental organizations (NGOs). We have referred repeatedly to their growing size and, in the West, political impact – particularly in those countries where green parties have acquired a foothold in the formal political system. In the US alone the 12 biggest NGOs have a combined membership of over 11 million people and a combined operating budget of more than $300 million per annum. There is no doubt that on a whole range of issues, from Antarctica through the North Sea to the ozone layer and climate change, the quality of their research and the energy of their lobbying, both in public and in private, have helped to mould public attitudes and influence government policy. (This is not to say they are always right; like other large organizations they have their blind spots and fixations, of which their doubts about free trade and aversion to nuclear energy, despite the evident need for at least a reconsideration of the issue in the light of global

---

[25] Thus trace gases in the atmosphere in 1974 were identifiable at a proportion of about one part in a million. They are now identifiable at one billionth of that.

The future

warming, are perhaps the most apparent.[26]) By the end of the Rio process they were omnipresent, they were included in most Western official delegations, and a great deal of their thinking found its way into the final documents.

This does not mean that NGO influence is decisive. The British government held out firmly against their insistence that the UK should join the 30% Club, and the US against their demand for carbon dioxide controls. However, on smaller issues, such as the Antarctic Minerals Convention, their pressure can make a large difference. The extent of this influence will no doubt now fall as public attention to the environment diminishes after the Rio climax (as it did after Stockholm), but it seems bound to remain substantial. International environmental discussions are uniquely distinguished from other types of international business by the presence and involvement of these large non-governmental pressure groups, not only in the negotiations themselves but also in the business of public agenda-setting and monitoring of agreements. Nor is their influence only political. Northern NGOs now distribute more funds in the developing countries than the World Bank.[27] The main point to make about their future influence on international environmental business is that, being heavily internationally oriented, some of them in their formal organization and some of them through fraternal links, they will continue to act as a powerful force for international cohesion on environmental matters, and cohesion oriented in a 'green' direction – a role we have already seen them play in all the big environmental negotiations of the late 1980s and 1990s.

> By the end of the Rio process NGOs were omnipresent, and a great deal of their thinking found its way into the final documents. Nor is their influence only political. Northern NGOs now distribute more funds in the developing countries than the World Bank. They will continue to act as a powerful force for international cohesion on environmental matters.

---

[26] Hostility to nuclear energy is, for example, the major flaw in the otherwise admirable and very balanced *Global Warming: The Greenpeace Report*, edited by J. Leggett.
[27] *World Resources 1992-3*, p218.

Environmental altruism

Our final factor for cohesion has been little remarked upon. In a striking number of environmental negotiations countries seem (despite Machiavelli) to have conceded more than a cold calculation of their national interests would suggest was justified. We have seen a number of examples of this among both developed and developing countries: British acceptance of the European Community directive on sulphur dioxide emissions, Japanese and European Community agreement to cut CFC production, Egyptian action to limit pollution of the Mediterranean, Mexican acceptance of the Montreal Protocol even before a special fund was created to help developing countries, Brazilian action to protect the rainforest, Thai adherence to CITES and so on. Even more striking examples are provided by Canadian and Russian support for action to tackle global warming, despite the possibility that global warming could substantially improve the agricultural productivity of both countries. There is also the very widespread phenomenon of states taking a unilateral lead on environmental issues in the hope, but not knowledge, that others will follow. The US ban on aerosols and the German cut of 25% in carbon dioxide emissions are conspicuous examples of this. (It is striking that we do not see similar unilateral gestures on, say, disarmament; and while we do see unilateral tariff cuts these are not comparable as they are of direct benefit to consumers in the country enacting them).

In each such case of apparent international altruism one can identify conjunctural political or economic factors which pushed the country concerned into line (international NGO pressure, British privatization of the electricity industry, a Mexican wish to be seen as one of the industrialized nations etc.). But the pattern of such concessions suggests that something more may be at work. In this context it is striking that, in the closely related area of international aid, altruism has been a key determinant of Western government policy.[28] One could tentatively hypothesize that something in the nature of the

---

[28] See D. Lumsdaine, *Moral Vision in International Politics*, Princeton University Press, Princeton 1993, for a persuasive elaboration of this point. In particular, Lumsdaine suggests that the evolution of international aid spending has to an extent been driven, independently of the self-interest of the donor governments, by its rhetorical justification – 'to help the poor'. It seems likely that a similar rhetorical constraint – 'to help the planet' – may operate in the case of international environmental policy.

environmental subject matter (such as precisely the feeling that one is contributing to the 'common good' and 'the welfare of future generations') tends to make politicians and negotiators readier to look for common ground than they would be in other sorts of international negotiation. This, of course, certainly does not mean that they will be willing to abandon their national interest, simply that they may be willing to bend it slightly more than they would in other circumstances to get an agreed outcome. Certainly something of that feeling emerges from the records of the chief US negotiator in the ozone layer negotiation,[29] and my personal experience of various EC negotiations as well as the climate change convention has given me something of the same impression.

11.6 The environment and the international system

Alongside the four forces for cohesion discussed above, the past twenty years have also seen environmental issues increasingly impinge upon existing international structures, notably the UN system, international funding arrangements and the global trade system. The adaptation of these structures to accommodate environmental concerns has raised a number of questions.

The UN system

Stockholm, of course, intended UNEP as the environmental focus of the UN system and UNEP has indeed developed over the past two decades from a marginal player to become the principal midwife of international action on new environmental problems as they arise – a role it successfully played on the Mediterranean, international toxic waste movements and biodiversity. It is thus now fulfilling half its original mandate as 'catalyst' for environmental action within the UN system. But the other half of its mandate, as 'coordinator', conspicuously failed with the death of the Environment Coordination Board in the 1970s. It is moreover difficult to be confident that role can be revived despite the calls at Rio for improved coordination. Not only do the big UN agencies remain as vigorously independent and mutually competitive

---

[29] R. Benedick, *Ozone Diplomacy*, Harvard, Cambridge Mass 1991.

as ever but recent years have also seen the emergence, as discussed above, of 'process oriented' environmental agreements in which states meet regularly to strengthen their commitments, and which have also given birth to a range of small secretariats to administer such agreements and organize the meetings. Thus there are separate secretariats for CITES, MARPOL, the ozone layer, climate change and so on. This rather untidy assortment of single-issue negotiations and secretariats has, however, been criticized as both inefficient and inadequate to enforce the agreements they administer.[30] In particular it has been suggested that they might be replaced by a unified body and forum for environmental negotiations on the model of GATT for trade issues.[31]

There are, however, persuasive arguments against such a change. The individual issue bodies have, broadly speaking, worked well. They cover a very heterogeneous set of problems, ranging from toxic waste to climate change, whose bundling together seems unlikely to yield substantial efficiency gains. They are able to pursue most of their work in a relatively low-profile and technical way, free from the political crosscurrents and over-bureaucratization which can affect the activities of larger organizations such as the big UN agencies and GATT. They are indeed too small effectively to enforce the environmental agreements they administer, and there are signs that some of the requirements of those agreements (notably on reporting) are being neglected.[32] This is certainly an argument for strengthening the secretariats; but there is no evidence that nation states wilfully breach their international environmental obligations any more than they do their other international obligations – and action to remedy significant breaches, in the environment as in other areas of international business, is not a matter for the secretariat so much as for a political response by other parties to the agreement. In the case

---

[30] H. French, *After the Earth Summit - The Future of Environmental Governance*, Worldwatch Institute, Washington DC 1992.

[31] Ibid.

[32] Thus the eight most important international environmental agreements to which the US was a party in 1992 (including the Montreal Protocol, CITES, MARPOL and the ITTO) had secretariats ranging from four to 20 people and suffered from significant problems of underreporting by parties and limited secretariat powers to verify implementation (United States Congress General Accounting Office report, 'International Environment; International Agreements are not Well Monitored', 1992). See also Hurrell and Kingsbury, *International Politics of the Environment*; Abram and Antonia Chayes, 'On Compliance', *International Organisation*, 47,2, Spring 1993.

The future

of the environment, an additional helpful factor is the NGOs, who already closely monitor national actions under the various agreements and can be expected to blow the whistle loudly if they detect any breach.[33] As Mohamed Al Ashry of the World Bank has persuasively put it:

> Although international institutions have not been systematically integrated their environmental efforts have complemented each other better than might have been expected. Their achievements stem not from large bureaucratic operations or enforcement powers, but from their catalytic role in increasing government concern, enhancing the contractual environment, and increasing national political and administrative capacity.[34]

International funding

Another area where a rather untidy pattern is emerging is on international funding for the environment. Here we have individual national aid programmes which, to a greater or lesser degree, have been won over to the doctrine of sustainable development (for example, the environment is now one of the central objectives of British overseas aid). Then there are the big multilateral programmes such as the UNDP and the World Bank, all of which declared at Rio that the environment was a leading priority for them, but, as we have seen, they had made similar declarations at Stockholm and had nevertheless financed some highly criticized projects in the interim. Some of this multilateral activity has been specifically structured for environmental ends, notably in the TFAP. Finally there is a small group of new, dedicated environmental instruments: the ozone layer fund, the Brazilian forest programme and the GEF.

It will be no easier (and perhaps no more desirable) to rationalize these arrangements than to rationalize the entire system for transfer of resources on concessional terms from North to South – with all of its inefficiencies, duplications and misallocation of funds for political purposes. A couple of points can nonetheless be made. The first is the likely future significance of the GEF. If, as seems likely, international action to deal with global environmen-

---

[33] Abram and Antonia Chayes, "On Compliance", *International Organisation*, Spring 1993.
[34] Quoted in P. Birnie; "The UN and the Environment", in *United Nations, Divided World*, Roberts and Kingsbury, 2nd edn, Oxford 1993.

tal issues intensifies in years to come then (subject to the solution of the problems of 'transparency and democracy' mentioned above) the GEF seems certain to become the central instrument by which to channel aid from the North to the South in connection with those problems. As we have seen, this could eventually involve sums of money amounting to a significant proportion of total global aid flows. Efficiency considerations (which operate much more powerfully here than in the case of individual subject secretariats because of the very specialized combination of financial and environmental expertise required as well as the possibility of 'multipurpose' projects) argue strongly in favour of managing these flows through a single fund rather than creating new instruments as new issues (such as desertification) come along. Indeed, the separate existence of the ozone layer fund, although perhaps politically difficult to terminate, makes no sense if an integrated alternative exists in which all parties have confidence.

The second point is the need for environmental vigilance over the activities of the World Bank and the other big multilateral aid agencies. Despite their protestations at Rio and their espousal of the doctrine of sustainable development, the primary orientation of these bodies is developmental, not environmental. The pressure exerted on them by recipient nations is for projects that boost growth, perhaps at substantial environmental cost. Only sustained pressure from the environmentally conscious West can ensure that they seriously assess the likely environmental impact of their projects before going ahead. The history of the subject suggests that, again, much of this vigilance will have to come from Western press and NGOs.

We have seen above, however, that far more of the money flowing from North to South comes from private rather than public sources. This is an aspect of future international environmental governance which was almost entirely ignored at Rio and continues to suffer from neglect. It is strongly arguable that the development of these private flows will have much more impact on international environmental problems in the future than the inevitably constrained official flows. We have already seen how Western private sector-transfers to developing countries (now running at about $80 billion per annum)[35] are likely to prove a channel, also, for Western environmental

---

[35] World Bank.

standards. We have also seen some innovative, if so far small-scale, ideas for the use of some of this money to help deal with global environmental problems, such as 'debt-for-nature swaps'. Another idea of the same kind, but with potentially much larger financial and environmental impact, was Norway's suggestion during the climate change negotiation that Western power companies should be allowed to achieve their emission cuts not through cutting back at home but through investments overseas, for example in afforestation or efficient power-generating technology, which would have the same effect on global carbon dioxide emissions. Even ahead of any financial benefit, some US power companies are already doing this in response to US public concern about climate change. The Norwegians estimate that this idea, on which work continues in the international climate change framework, could lead to private transfers of tens of billions of dollars to developing countries.

Trade

The other source of private financial flows to the developing world is of course through trade, and the relationship of trade to the environment is now a very hot topic.[36] The discussion at both theoretical and policy levels occasionally gives the impression of two rival priesthoods criticizing each other's creeds. Environmentalists argue that free trade, for example, in tropical timber and hazardous waste is directly damaging to the environment, and that trade encourages the exploitation of developing country resources and the spread of industry (some of it highly polluting) to the laxer environmental regimes often to be found in developing countries. More generally they point out that 'more trade tends to mean more economic growth....increased growth will lead to increased energy use, materials consumption and therefore pollution and waste generation'.[37] And they argue that inadequate environmental legislation amounts to a form of subsidization of that country's economic production, and should be treated as such in international trade agreements such as the GATT. Trade experts, on the other hand, emphasize the prosperity that free trade has brought (including, increasingly, to developing coun-

---

[36] See e.g. P.Low, ed., 'International Trade and the Environment', World Bank Discussion Paper, 1992.
[37] H. French, 'Reconciling Trade and the Environment', in Brown et al., *State of the World 1993*, Earthscan, London 1993.

tries). They also underline the danger of allowing the environment to become a catch-all excuse for restrictions – which would be only too easy if the concept of 'environmental subsidization' were given free play.

There is no doubt that trade links have played a major role in the evolution of international environmental protection. We have seen the prevalence of the 'level playing field' effect – the concern for the competitivity of their domestic industry which has regularly led countries to try to get their domestic environmental standards followed internationally. We have also seen the direct use of trade controls for environmental ends in CITES and the Montreal Protocol, as well as occasional unilateral trade threats by the US (and the EC has occasionally acted similarly) to compel other countries to take environmental action. It seems very likely that, given intense Western public environmental concern, this sort of action will grow, through the agency of both governments (the Austrian government already requires tropical timber to be specially labelled, and there are pressures for a similar requirement in the EC) and private companies (as with B&Q's refusal to buy unsustainably forested timber).

These particular instances, however, should not be allowed to obscure a central truth. The world free trade regime is one of the most effective instruments we have to help raise global environmental standards. Countries are in any case going to work to achieve economic growth, and in the case of the developing countries we have seen that growth is the route most likely in time to supply the resources and will to tackle environmental degradation. To criticize the environmental impact of free trade on the grounds that it promotes economic growth therefore misses the point (quite apart from coming close to the morally dubious and politically impracticable view that the way to protect the global environment is to keep the poor poor). Given that growth will take place anyway, the question is whether under free trade it will be more, or less, environmentally friendly. The answer is plainly that it will be more so. Under free trade, goods are produced where it is most efficient to do so, rather than where subsidies and duties push them. Thus under free trade a given level of world product will consume less resources and produce less pollution than when trade is distorted. We have already seen a compelling example of this in the case of the world's agricultural market. A similar case has been made for the world's energy market, where free trade in coal combined with abandonment of the West's $60 billion per annum subsidies to its coal producers would, in a manner which brought significant economic benefit to all, also bring a cut in world consumption of coal and so in carbon dioxide emissions.[38] The introduction of trade controls for environmental reasons does indeed offer vast scope for protectionism: if, for example, countries introducing carbon taxes were free also to impose tariffs on the exports of countries which they claimed were doing less to tackle climate change, such action could in principle affect the entire range of a country's output.

This does not mean that the current global trade regime is perfect from the environmental point of view. The 'environmental subsidization' argument has some force, at least as applied to *international* environmental problems.[39] There are plainly goods whose production has global environmental costs which the producing country may wish to exclude from their sale price in the hope of boosting exports. The most likely example in the immediate future of this form of 'environmental dumping' is goods whose production will produce significant greenhouse gas emissions into the atmosphere. There may well be a case on grounds of global economic efficiency and environmental health for the international community to impose a penalty on such goods corresponding to their global environmental cost. Such a scheme would correspond to a more sophisticated version of the Montreal Protocol and CITES arrangements for penalising the trade of those who are not contributing to the global environmental effort (which in both cases have played a significant role in encouraging countries to join that effort). The important point is that it should not be open to individual states to impose such penalties: that way lies trade chaos and protectionism. As with the Montreal Protocol and CITES

---

[38] K. Anderson: 'Effects on the Environment and Welfare of Liberalising World Trade, the Cases of Coal and Food', in Anderson and Blackhurst, *The Greening of World Trade Issues*, Harvester Wheatsheaf, London 1992.

[39] Although French, in 'Reconciling Trade and the Environment', goes a bit far with her suggestion that there should be a set of 'common environmental standards' which, if not met, should justify restraints on that country's trade. The problem with this is that, applied to purely domestic environmental consequences of production, it could amount to the environmental equivalent of penalizing a country's exports because eg it pays low wages – which is plainly nonsense as applied to poor countries. The question is whether a country's environmental failings impose environmental costs on other countries. If so (and climate change is the most plausible example) then penalization may be justified.

the rules should be set multilaterally, for instance by those taking serious action together to tackle climate change – which will not necessarily include all signatories to the Climate Change Convention.

A political centrepiece?

Thus, through evolution of some new components (such as the individual treaty secretariats, the GEF and UNEP) and adjustment of others (such as the UN agencies and the international funding and trade structures) the international system is gradually responding to the new environmental requirements being placed upon it. There was nevertheless a strong feeling at Rio, as there had been at Stockholm before, that some kind of a political centrepiece was needed. The Stockholm answer was UNEP, which spent years languishing on the fringes before finding a role for itself. The Rio answer was the Commission for Sustainable Development, which is as yet untested. Its aim is to keep the environment/development debate and the pursuit of sustainable development in the public eye by discussing these issues, at high level, annually. It is difficult to be optimistic about this process which seems dangerously likely to reproduce the ritualistic and cloistered annual exchanges which already take place on comparable issues in the UN General Assembly and its Economic and Social Council.

Is there any need, then, other than to flesh out the output of UN environment conferences, for any kind of central institution? An answer is given by the experience of Stockholm and (probably) Rio themselves. Both these conferences had the effect of pushing the environment up the agendas of nation states and multilateral organizations. As a result of Stockholm, states created environment ministries and began legislating on environmental issues. Multilateral financial institutions for a while took the environment seriously into account in planning projects. The effects faded over time, partly because the work of environmental repair was in hand and partly because political attention moved elsewhere. Rio has however renewed and reinforced the impact of Stockholm; and the effects of Rio can be expected in due course to fade in the same way. This suggests that there is a case for a global high-level 'attention-centring' gathering every so often to keep the subject in a prominent position on national and international agendas. If such meetings take place too often they become merely routine, with ministers often not bothering even to show up. But it is arguable that 20 years is too long to wait for a conference to follow up Rio.

One other sort of central environmental institution has been proposed in the form of an 'Environmental Security Council' – a supreme UN body composed of a restricted number of states to take up topical environmental issues (perhaps formed by using the Security Council itself, adapting the UN Trusteeship Council or setting up a special body of similar status[40]). Some caution is required here. The UN Security Council deals with political or military emergencies. Environmental issues rarely come in the form of emergencies, but rather as slow moving, complex problems which require a great deal of technical background work if they are to be handled correctly. The system of single-issue negotiating structures which has emerged to deal with many of them seems successful. It would be a mistake to suck the effectiveness out of such structures, and introduce the possibility of extraneous politicization and stalemate which have so often been a feature of work in the UN, by creating some grand central council over their heads. That said, however, it is clear that a group of about 20 states (the G7, Russia and the ten or so largest industrializing countries) have been at the heart of every global environmental negotiation. Regular meetings within this group (which might be extended, via a constituency system, to include representatives of the small countries while still keeping the body to a relatively manageable size) might provide opportunities for an informal overview of international environmental business, and the avoidance of significant misunderstandings and splits. Such meetings could be no more than consultative. The group would certainly not be able (on the Security Council model) to take binding decisions on individual issues. The environment is far too pervasive a subject for those excluded to permit this, as the ill-fated Hague Declaration showed. The smaller group could be no substitute for the formal proceedings of the individual negotiating bodies, which would remain sovereign over their own subjects; but it could help to introduce some coherence and sense of shared direction into the fast-growing number of international environmental institutions and activities.

---

[40] Cf. J.MacNiell, P. Winsemius and T. Yakushiji, *Beyond Interdependence*, Oxford University Press, Oxford 1991.

11.7 Envoi: back to the roots

I would not want to conclude with the impression that solving the world's environmental problems is finally a matter of creating the right international institutions and rules. Environmentalism, with its emphasis on humanity's growing collective impact on the state of the planet, and its instinct that 'everything is connected to everything else',[41] has fallen rather too readily into the view that the solution is to be found through dense webs of regulation and grand schemes of national and international planning (sometimes, as we have seen, with the suppression of individual liberties as an accompaniment).[42]

The fact is that the replacement of the judgement of the individual by that of the state raises problems of its own. Governments are swayed by powerful special interests and political cross currents which divert them, often at extremely high cost, from optimal policies. We have seen a number of examples in the environmental field where it has been precisely the excessive role of the state that has produced destructive consequences, notably in Eastern Europe and the shape of world agriculture. The transfer of decision-making authority a further step up to some international body, even if it were politically achievable (which the fate of the Hague Declaration and the highly voluntary nature of most existing international environmental institutions suggest it is not), would have the effect of placing it still further from the people it is intended to serve, and making it still more subject to political manipulation, as much of the history of the UN demonstrates.

These difficulties associated with looking to a centralization of authority to bring about environmental progress cast into some relief that quieter component of the growth of international environmental action over the past 20 years which has lain not in the grandiloquent international declarations and the painfully negotiated treaties, but rather in the gradual dissemination to more and more levels of national and international society of awareness of,

---

[41] This is one of the 'laws of ecology' set out in Barry Commoner's *The Closing Circle: Nature, Man and Technology*, Knopf, New York 1971.

[42] It is striking, especially since the demise of the ideal of the planned economy in the late 1980s, how many authors now offer the environment as a major justification for their doubts about free market economics. See e.g. P. Keegan, *The Spectre of Capitalism*, Radius, London 1992, and J. Dunn, *Western Political Theory in the Face of the Future*, 2nd edn, Cambridge University Press, Cambridge 1993.

and responsibility for, the costs of environmental destruction. As a result of this process individuals, private organizations and individual nation-state governments find themselves increasingly in the position where the actions from which they themselves benefit are also those which serve the greater global good. We have seen numerous examples of this in the preceding pages. Thus in the case of population policy the complete sterility of global discussion of the subject has contrasted strikingly with the growing proportion of the human race which is deciding (given the relevant information and incentives) to keep their family size down as they become aware of the personal costs to them of the alternative. Similarly, we have seen a growing proportion of Western industry beginning to take the environment seriously under pressure from an aroused public and the danger of escalating liability costs. On species, the 'ranching' clauses of CITES, the Merck/Costa Rica deal and the development of ecological tourism all point to ways in which countries are beginning to find value in conserving their own biodiversity. The same is likely to become the case for forests, as forest-owning countries begin to recognize (where necessary with Western scientific help) the full economic value of their forests to them in terms of ecotourism, forest products, preventing soil erosion, or whatever.

On finance, too, it seems likely that the fruitless argument about official aid levels at Rio missed the point. The booming area of international transfers is in the private sector, through both trade and international investment. Developing countries will participate in the gains not by arguing at international conferences, but by running their economies well. We have also seen that these flows of private funds tend both to bring Western environmental standards directly and, by boosting the prosperity of developing countries, to help them towards a greater capacity and willingness to tackle environmental degradation. It is also through the private sector that the most innovative thinking on spending for the environment is currently taking place, for example, in the form of 'debt-for-nature' swaps and the Norwegian ideas on private financing for carbon dioxide reduction schemes.

This does not mean that there is no role for international governmental action. The plight of the least developed is not going to be solved by the private sector. It requires (a rare phenomenon in international affairs) an act of imaginative generosity by Western governments. Climate change is plainly a problem which, in the tradition of the 'commons' problems of the past – the oceans, acid rain, the ozone layer – needs to be tackled through a process of international agreement. But even this is an issue whose potential costs are so high that failure to exploit the efficiencies generated by use of private incentives could easily wreck the whole enterprise – whence the importance of Norwegian ideas on private financing, the widespread current discussion of carbon taxes and tradable emissions rights, and the emphasis on finding economically attractive 'no regrets' measures to cut greenhouse gas emissions.

> **The Rio Conference may be viewed as one of the latter gasps of the old 'command and control' style of environmental thinking ... The key failures at Rio may turn out to have been not a setback but a necessary impulse to a new philosophy of global environmental management which is still struggling to be born.**

The broad implications of this approach are a shift away from the 'international planning' model for resolution of the problems of the planet, and recognition that the role of the international process is likely to be much more the setting of general directions and incentives, and the exchange of information and modest amounts of help, than the imposing of rules on nation states (even when that is politically possible) and the pressuring of environmental recalcitrants into line. It is striking that in tackling their own domestic environmental problems governments are increasingly using the same technique of setting financial incentives rather than introducing prescriptive regulation (as with emission trading arrangements in the US, water charging in China, and petrol taxes in Europe). And the huge environmental and economic benefits to be gained simply by eliminating the vast global morass of economic subsidies and incentives which are directly damaging to the environment (whether ranching in the Amazon or agricultural overproduction in Western Europe) are beginning to be identified and reaped, notably in connection with western action to tackle global warming.

In this light, the Rio Conference may be viewed as one of the latter gasps of the old 'command and control' style of environmental thinking with, on the one hand, Western pressure for global statements on population and conventions on forests and biodiversity and, on the other, Southern demands for a desertification convention and huge quantities of official Western aid. It seems likely that most of these subjects could be more effectively addressed not by their aggregation into global packages, but by their disaggregation into individual national, or smaller, units. The key failures of Rio may turn out to have been not a setback but a necessary impulse to a new philosophy of global environmental management which is still struggling to be born.

This shift in emphasis towards the private and the local, however partial and tentative it has so far been, is perhaps the most significant change of course in recent environmental affairs. It offers for the environment what free markets offer for the economy. It places the responsibility, and incentive, to act with individuals, companies and national governments rather than some remote (and almost certainly over-politicized and underfunded) international agency. It has the important democratic benefit, over a period when the growing scale of human economic activity is going to require a comparable growth in action to protect the environment, of ensuring that action is carried out in a way which leaves the maximum range of choice to individuals and private organizations, and minimizes concentrations of economic and bureaucratic power. It does not mean that there will be no more need for international environmental rules and agreements – only that those agreements will increasingly work with the grain of market economics and individual incentives rather than across it. If they are framed correctly the incentive structure will be such that nation states will protect the global environment through the pursuit of their own selfish national interests – as they are already beginning to do on such issues as population policy and energy pricing. Machiavelli would have approved.

Index

acid rain 3, 104–7
  Canada 105, 165–6
  European Community 110–11
  increase 55
  legislation 11
  Swedish concern 34, 41, 104–5
  UK attitude 146
Africa
  aid requirements 246–7
  climate change policy 180, 181
  food production 56, 58, 76
  forests 78, 152
  ivory trade 16, 101, 103
  population 58, 64
  toxic waste 127, 131–2
'Agenda 21' 11, 161, 211, 213–14, 220, 228
  cost 220, 230
agriculture 3, 52, 75–83, 86
  in developing countries 58–9, 69, 179–80
  fertilizers 52, 76–7, 78, 81, 244
aid, international
  agencies 262
  for 'Agenda 21' 220–1, 230
  to Brazil 155, 156
  linked to environmental projects 38, 47, 120, 209, 212, 261
    CFC–linked 142–3
    climate change 190
    for conservation 200
    population projects 122
  1987 plea 128
  Lomé Conventions 120–1, 132–3
  national proportions given 212
  Stockholm Conference 39, 46
  see also Global Environment Fund
air pollution *see* atmospheric pollution

air transport 2, 41, 135
Al Ashry, Mohamed 261
Algeria 97, 100
Alkali Act, *1863* 15
Alliance of Small Island States (AOSIS) 180n, 181, 184
altruism 258–9
Amazonia 9, 67, 152, 154–6
  Amazon Pact Treaty 107
animals *see* biodiversity; conservation
Antarctica 133–4, 212
  mineral reserves 133–4, 257
  ozone layer 134, 138–45
  treaties 90
Apollo II 23
Aral Sea 72
Argentina 45, 61
Arrhenius, Svante 163
asbestos 243
atmospheric pollution 104–7
  developing countries 59–60
  Mexico 30
  OECD 53
  treaties 89, 90
    Long Range Convention 73–4, 105–6, 137, 138
  UK 11, 21
  UNCED 214
  US 104, 214
  see also acid rain; greenhouse gases
Audubon Society, National 19
Australia 134, 189

Baker, James 170, 172
Baltic Sea 92
Bangkok 60

Bangladesh 127, 179, 239
Basle Convention on the Transboundary Movements of Hazardous Waste 132–3
Belgrade Summit 219–20
Bellagio workshop 165
Bergen, Norway, *1990* meeting 172–3
Bhopal, India, disaster at 56, 127, 146
biodiversity Convention 161, 197–200, 269
  negotiation 201–4
  aftermath 204–6, 208
  future 248–9
biosphere 200
birds, protection 101
  migrating 15, 16
  oil 17, 18
Bonn Convention 92
Boumedienne, President of Algeria 97
Brandt Report 120
Brazil
  agriculture 81
  climate change policy 180, 189
  Cubatao 86n
  dam project 45
  development 38–9, 180
  environmental concern 67, 86n
  forests 9, 41, 127–8, 151–2, 154–7, 159, 201, 249
  UN conference hosted 160, 215
  waters, coastal 42
British Central Electricity Generating Board 146
British Nuclear Fuels Ltd 148
Brundtland, Gro Harlem 128–9, 166, 246
Brundtland Report, *Our Common Future* 75n, 122, 128–9, 160, 208, 219, 238
Bucharest
  conference on population 63
Budapest 72
Bulgaria 66, 74
Bush, President George 129–30, 166–7, 170, 172, 190, 191, 205
  at UNCED 224, 226, 227, 229, 234

cadmium pollution 22
Caldwell, L. 24n, 84
Canada
  acid rain 105
  CFCs policy 165–6
  coastal waters 41
  conservation 16
  greenhouse gases 189
  see also Montreal Protocol

Cancún Summit 121
carbon dioxide (CO_2) 52
  see also greenhouse gases
cars 55, 56–7, 146, 150
  petrol 54
Carson, Rachel 18–19, 25
Catholic church 61–2, 64, 214
CFCs *see* chlorofluorocarbons
chemical companies 149, 248
  CFCs use 135–6, 141, 143, 144–5
  drugs production 198n, 199, 204–5
Chernobyl, Ukraine, disaster 2, 3, 56, 130–1
China
  agriculture 81
  CFCs 142, 143
  climate change policy 179, 180, 187, 188, 189
  environmental concern 65, 68–9, 70, 245
  pollution 65, 71
  population 40–1, 61, 62, 63, 86
  Tiananmen Square 2
chlorine catastrophe theory 137
chlorofluorocarbons (CFCs)
  as greenhouse gas 163
  ozone layer effect 135
  reductions policy 135–45, 146, 150, 167
    developing countries 142–3
    Montreal Protocol 140–3
    Vienna Convention 137–8
CITES *see* Convention on International Trade in Endangered Species
cities 59
Clean Air Acts
  UK 11, 21
  US 104
climate change 3–4, 127, 157–8, 163–95
  CO_2 emissions reduction targets 173–7
  concern increase 165–7
  conferences on: *1979* 165; *1990* 173, 176, 183–5, 186
  Convention 161, 185–91, 232
    negotiations 193–5
    established 191–3, 195
  developing countries policy 179–83, 184–5
  early work on 163–5
  effects 166–7
  future 250–2, 270
  governmental response 167–71
  Intergovernmental Negotiating Committee 185–91
  IPCC established 177–9
  pressure on US policy 171–3

Index

UNCED preparation 208
  see also Intergovernmental Panel on Climate Change
climatic disasters 127, 166
Clinton, President Bill 193, 205
Closing Circle, The (Commoner) 27
Club of Rome 27, 29, 52, 75, 120, 240
coal 52, 265
Collor, Fernando, President of Brazil 155–6
Commoner, Barry 27
commons, tragedy of the 4
communications and media 2, 23, 55–6, 247

communist bloc 30, 36–7, 71–5
Concorde 135
conferences, UN 35–6
  see also names of conferences
conservation 15, 16, 122
  biodiversity 161, 197–206, 248–9, 269
  birds 101
  fish 119, 212
  treaties 90
  wetlands 31
  wildlife 16, 31
    CITES 100–4, 197, 265
    in tropical forests 151
  see also forests
consumerism 148–9
consumption 3–4, 51–2, 240n
Convention on International Trade in Endangered Species (CITES) 100–4, 197, 265
Convention on the Regulation of Antarctic Mineral Resource Activities (CRAMRA) 134
Costa Rica 248
counter-culture 22–3
CRISTAL 32, 147
crocodiles 101, 102
currencies 2
CSD (Commission on Sustainable Development) 219, 228, 266
Czechoslovakia 72, 74

Dahl, R. 84n
Daily Telegraph 29
Daly, H. 240n
Delors, Jacques 205
democracy 30, 241–5, 255–6
Denmark 113, 126
desertification 80

Conference on 121, 123
  at UNCED 212–13, 215, 216, 229
developing countries
  agriculture 77–83
  CFCs 142
  conservation 101, 102, 200, 201–4
  development and industrialization 59–60, 61, 64–71, 142, 179, 218 10
  environmental concern 30–1, 70–1, 123, 244–7, 247–8
  spending 70
  forests policy 41, 151–7, 215–16, 229, 249
  GEF, attitude to 188, 209–10
  government control 8–9, 11, 244
  greenhouse gas emissions 173, 179–83, 184–5, 186–7, 188, 191, 194–5
  G77 187, 188–9, 225
  marine pollution 96
  population problems 57–9, 60–4
  problems 57–60
  Stockholm conference preparations 37–42, 44–6
  technology for 142–3, 219–20, 228–9
  toxic waste 131–3
  UNCED objectives 212–21, 225–6
  see also aid, international; desertification; North–South divergence; names of countries
development 10
  sustainable 172–3
  Commission on Sustainable Development see CSD
  see also industrialization
disasters, environmental 127–8, 146, 166
  Chernobyl, Ukraine 2, 3, 56, 130–1
  Exxon Valdez 146, 148, 149, 150
Doomsday Syndrome, The (Maddox) 29
drugs production 198n, 199, 204–5
Dubos, Rene 36
Dupont Corporation 135, 139, 141, 144, 149

'Earth Charter' see Rio Declaration
'Earth Day', *1970* 19, 36
Eastern Europe 71–5
  climate change policy 183
  see also communist bloc; names of countries
ECO newspaper 43
ecology, term 15
economic growth 10, 51–3, 58–60, 264–5
  limits 240
economic order, international 119–20

Economist 5, 149–50
EcoTech fair 224
Ehrlich, Paul 27, 29, 41n, 43, 62
energy
  conference on sources 121
  consumption 51, 240n
  efficiency 250
  industry, CO_2 emissions attitude 167–8
  market 265
  nuclear 115–18, 256–7
  production industry 51, 167–8
  see also fossil fuels
environment, term 15, 18–20
Environment Coordination Board 49, 50, 217, 259
Environment Defense Fund 19
Environment Fund, UNEP 48, 208
Environmental Protection Agency (EPA) 147, 167
  Science Advisory Board 243
Ethiopia 127, 239
European Community 23, 239–40
  aid donations 230
  CFCs 139–40, 144
  CO_2 emissions policy 167, 175–6, 184–5, 186, 192–3
  environmental concern 126
  environmental legislation 11, 54, 90, 99, 106–7, 109–14, 256
  forests policy 229
  industry 146
  nuclear power 117–18, 130–1
  toxic waste disposal 132–3
  treaties 114–15
  UNCED 234–5
European Parliament 126
Exxon Valdez 146, 148, 149, 150

famine 56, 127, 247
fertilizers 52, 76–7, 79, 81, 244
fish conservation 97, 119, 212
Foley, G. 240n
food
  production 52, 56, 58, 76
  famine 56, 127, 247
  see also agriculture
  radioactivity levels 131, 243, 254
Food and Agriculture Organization (FAO) 16, 47, 50
  Mediterranean Action Plan 97
Foreign Affairs 29
forests 52–3, 269
  tropical 3, 9, 41, 67, 151–7
    aid for 153, 155
    biodiversity 151, 152, 198, 201–4
    future 248–9
    greenhouse gases 151, 186
    UNCED 161, 215–16, 229–30
fossil fuels
  developing countries 248
  pollution effect 21, 163, 165, 166, 167, 169, 182, 183
  resources 52
Founex Report 37–8, 39
Fourier, Jean Baptiste 163
France
  acid rain 106
  Antarctica policy 134
  conservation policy 204
  environmental concern 126, 158, 159
  environmental legislation 54
    resistance 98, 99, 146
  Mediterranean pollution 99
  nuclear power 116–17, 118, 169
freedom 84–7
  of the seas 41
French, H. 265n
Friends of the Earth 19, 126
funding, international 261–3, 269–70
  see also aid, international; Global Environment Facility
future, the 237–71

G7 159, 172
G77 187, 188–9, 225
Gandhi, Indira 37, 41, 42, 68
GATT 8, 90, 142, 260, 263
Geneva meetings
  1979 165; *1988* 168; *1991* 187–9
Germany
  acid rain 106, 110–11
  CFCs 140
  CO_2 emissions 174
  East 37
  environmental legislation 54
  nuclear power 131
Global Environment Facility (GEF) 161, 188, 189, 192, 208–10, 261–2
  for biodiversity 203, 204
  UNCED 220
Global Forum 224
Global 2000 Report to the President of the United States 120

Index

global warming 163–7, 257
Gorbachev, President Mikhail 158
governments
  environmental concerns 6, 158–60, 242–4
  powers 2, 7–9
  treaties 10–12
  see also states, nation
Greece
  acid rain 111
  ancient 60
  1981 elections 66
Green Consumer Guide 148
'green' environmentalism
  consumers 148–9
  'dark' 28–9, 84–5, 239
  'light' 28
Green Parties 26, 56, 126, 239
  France 126
  Germany 26, 140
  New Zealand 20, 56
  Sweden 126
  UK 126
greenhouse gases 151, 163–72, 177, 178, 183–6
  effects 163–5
  IPCC work 177–9
  restriction policy 166–71, 173–7
    Convention 185–90, 191–5
    developing countries 179–83, 194–5
    US, pressure on 171–3
  tropical forests 151, 186
  see also climate change
Greenpeace 19, 126
growth, economic 10, 51, 239–40, 264–5
  limits 28, 240
  sustainable development
Gulf War 56, 212

Haas, P. 255n
Hague Declaration 159, 267, 268
Hansen, James 166, 170, 255
Harding, Garret 4
Harrison, P. 249n
heavy metals, toxic 21
Heilbronner, R. 84
Houghton, John 177
Houston, G7 meeting *1990* 172–3, 215
Hugo, Victor 15
Humanae Vitae 61
Hungary 74
hurricanes 166
Huxley, Aldous 15

India 37
  aid requests 128
  CFCs 142, 143
  climate change policy 180, 181, 187, 188, 189, 190
  environmental concern 65–6, 68, 70, 160, 245
  forests 151, 229
  government 9, 70
  industrial projects 30, 64–5
  land cultivation 81
  population 40–1, 61, 86
  UNCED 229
industrialization 37–9, 111
  in developing countries 59–60, 61, 64–71, 142, 179, 218
  Eastern Europe 72–3, 74
industries 13, 46–50
  CFCs use 139–43, 144–6
  OECD 53–7
  US 33, 135–6, 140, 142–3, 144
institutions, environmental 16, 217–19, 259–61, 268
Intergovernmental Negotiating Committee on climate change (INC) 185–91
Intergovernmental Panel on Climate Change (IPCC) 168–70, 177–9, 184, 186
  Bush addresses 172
  developing countries 180–2
  Working Group I 168, 177, 193–4, 254
  Working Group II 168, 177–8
  Working Group III 168, 170, 179
international agenda 1–2, 6
International Atomic Energy Agency (IAEA) 47, 48n, 118, 130
international conferences 12–13
  see also names of conferences
International Convention on Oil Pollution 31, 92, 93
International Convention on the Prevention of Pollution from Ships (MARPOL) 17–18, 92, 93–4
International Labour Organization (ILO) 16
International Maritime Organization (IMO) 16, 17, 91, 94–5
International Planned Parenthood Federation (IPPF) 64
International Tropical Timber Organization (ITTO) 153–4
International Union for the Conservation of Nature (IUCN) 100, 101, 122, 200

International Whaling Commission 16, 42, 212, 243
investment 31, 148, 149, 246
Ireland
  industrialization 31, 111, 189
  nuclear issues 118
'issue/attention cycle' 24, 55
Itai Itai disease 22
Italy 98, 99, 126, 146
  dioxin leak 55
Ivory Coast 39
ivory trade 16, 101, 103

Japan
  aid donations 230
  CFCs 136, 140, 144
  CO_2 emissions 169, 171–2, 174–5
  environmental concern 20, 24, 26, 56, 126, 174–5, 243
  growth 22
  industry 57, 148
  legislation 30, 54
  nuclear power 117
  pollution 22, 23
  production 244n
  UNCED 220, 221
  wildlife 102

Karin B (waste ship) 127, 131–2
Kenya 65–6, 103
Koko, Nigeria 131
Korea 117

Labour Organization, International 16
Labour Party 26
Lake Baikal 72, 73
land, cultivable
  in developing countries 58
  see also agriculture; forests
Large Combustion Plants Directive 107, 110–11
Latin America 64, 67, 202
lead 53, 54, 59, 60
legislation 10–12, 30, 243
  European Community 11, 54, 90, 99, 106–7, 109–14, 256
  OECD 53–4
Levin, Bernard 22–3
Lomé Conventions 120–1, 132–3
London: smog 5, 21
London Dumping Convention 92, 93, 94, 118
London meeting, *1990* 141, 142–3, 180–1

Long Range Transboundary Air Pollution Convention (LRTAP) 73–4, 105–6, 137, 138
Lumsdaine, D. 258n
Luxembourg meeting, *1990* 176

Machiavelli, Nicolò 1, 5, 6, 51
Maddox, John 5, 29, 52
Malaysia
  biodiversity attitude 204
  environmental legislation 67, 70
  forests policy 152, 153, 157, 215–16, 229
  UNCED 227, 229
Maldives meeting, *1990* 180
Malthus, Thomas 5, 29, 62
marine pollution
  coastal waters 41–2, 146
  Conference on the Law of the Sea 119
  Mediterranean 96–100
    land–based 98–100
  oil spills 16–17, 18, 31, 32, 94, 146, 147
  Oslo Convention 33
  treaties 17–18, 33, 89, 90, 91–5 (*listed* 92), 96
  regional 95–100
  waste dumping 32–3, 42, 98–9
    nuclear 21, 118
MARPOL (International Convention on the Prevention of Pollution from Ships) 17–18, 92, 93–4
media coverage 2, 23–4, 55–6, 247
Mediterranean Sea, pollution control 146
  Action Plan 92, 96–8
  Land–based Sources Protocol 92, 98–100
Melsiuki Convention 92
Mendes, Chico 127, 155, 156
mercury 22, 23
methane 163, 179, 181n
Mexico
  coastal waters 42
  environmental concern 30, 66, 70
  industrialization 9
  population 62
  rivers 59
  wildlife conservation 102n
Mexico City 30, 59, 60, 67
  conference on population 64
migration 2
minerals
  Antarctica 133–4
  mercury 22, 23
  reserves 52
  under–sea 119

Index

Mitsubishi 148
Monsanto 149
Montedison Company 98, 146
Montreal Protocol 140–4, 148, 265

Nairobi
  meetings: *1977* 121; *1981* 121; *1982* 121; *1991* 189; *1992* 203–4
  UNEP site 49
National Academy of Sciences, US 172
National Geographic magazine 127
Natural Resources Defense Council 19
Nature 5
Nature Protection and Wildlife Conservation Convention 16
  negotiating processes 252–4
Netherlands 169, 171, 174, 175, 176
New Delhi meeting, *1989* 180
New International Economic Order (NIEO) 119–20, 121, 123, 219
New York meeting, *1990* 190
New Zealand
  Green Party 20, 56
  whaling 212
Nicaragua 77
Nigeria 127, 131, 132
nitrogen oxide 52, 55
  in acid rain 104–7
  from aircraft 135
non–governmental organizations (NGOs) 3, 7–8, 256–8, 261
  in developing countries 66n
  Stockholm conference 43–4
  US 256
Noordwick, Holland, *1989* meeting 171, 180
Nordhaus, W. 251n
Nordic Convention on Protection of the Environment 107
North Sea 92, 94, 95, 243
North–South divergence 9–10
  re 'Agenda 21' 213–14
  re biodiversity 202–4, 214
  Brandt Report 120
  CFCs 142–145
  re Earth Charter 211–12, 214–15
  population 40–1, 43, 44–5, 63–4
  Stockholm conference 39–41, 44–5, 63
  after Stockholm 118–23
  UNCED 231
  see also developing countries
Norway 189, 263

environmental concern 104–5, 126
1990 meeting 172–3
nuclear issues 115–18
  accidents 2, 3, 56, 116, 130–1
  energy 115–18, 256–7
  waste 3, 21, 115–16, 117–18
  weapons 121
  protests 25
  testing 41, 118
  treaties 18, 90, 118
  see also International Atomic Energy Agency (IAEA)

Ocean and Coastal Areas Programme 96
oceans *see* marine pollution; seas and oceans
OECD
  environmental concern 53–7
  spending 70
  greenhouse gases policy 189–90, 191
oil pollution, marine 16–17, 18, 22, 31, 32, 55, 94, 146, 147
  treaties 32, 92
  Convention for Prevention 17–18, 31, 92, 93
oil–producing states 214
  attitude to climate change 182–3
oil reserves 52
Only One Earth (Ward and Dubos) 36
OPEC 183n
Ophuls, W. 84, 85
Oregon, rivers 21, 23
Organization for African Unity 131
organizations, international 16, 217–19, 259–61, 268
Oslo Convention 33, 92, 93, 94
Ostrom, E.E. 4n
Our Common Future see Brundtland Report
ozone layer 41, 127, 134–45, 146, 150
  background 134–7
  Montreal Protocol 140–4, 148, 265
  negotiation process 252–3
  Vienna Convention 137–8

Palme, Olaf 42
Paris, *1991* meeting 189–90
Paris Convention 92, 94–5, 98
Partial Nuclear Test Ban Treaty 18, 118
pesticides 18–19, 77–8, 79
petrol 54
Peugeot car company 146
Poland 72, 74, 132

pollution
  control 16, 53–5
  growth 20–2
  international spread 3, 10, 22
  response to 5, 15, 16–17, 18–20, 26–7, 55–7, 240–1
  see also atmospheric pollution; marine pollution; oil pollution
population
  conferences on 40, 63–4
  control policy 61–2, 86, 214, 269
    developing countries 40–1, 57–9, 60–4
    Stockholm Conference 40–1, 43, 44–5
    UN fund 63, 64
  world growth 3–4, 21, 51, 62, 240n
Population Bomb, The (Ehrlich) 27
Porritt, Jonathan 4, 28
Preservation of Animals ... in Africa, Convention 16
pressure groups 19, 26, 43–4, 243–4
production 20–1, 51–2, 244n, 245
prosperity 22, 241–4, 246
protests, mass 25

Ramsar Convention on Conservation 31
Reagan, President Ronald 57, 129–30, 158
recycling 54
refugees 2
regional agreements 89
Regional Seas Programme 95–100
Reilly, William 226–7, 243
resources, natural 3–4, 10, 16, 51–3, 240n
Rhine, river 16, 53, 56, 127
Rio Conference *see* United Nations Conference on Environment and Development (UNCED)
Rio Declaration ('Earth Charter') 161, 211, 214–15, 227–8
Ripert, M. 185, 191–2
rivers, pollution 21, 23, 53
  developing countries 58, 59
  Eastern Europe 72
  Rhine 16, 53, 56, 127
Romania 63, 74
Russia 73, 85n
  climate change policy 182, 183, 186
  see also Soviet Union

Sarawak 153–4, 157
Sarney, President of Brazil 155

Saudi Arabia
  CO_2 emissions policy 182–3, 184, 189, 190, 192, 193
Savasin, Paul 101
science and scientists 6, 27
  climate change 172
  Mediterranean Sea 97–8
  nuclear power 115
  ozone layer 137, 141, 144, 172, 254–6
seals 127
  fur 16
seas and oceans
  conference on law of 119, 123
  level, rise 180
  minerals beneath 119
  see also marine pollution; Mediterranean Sea
Sellafield 148
Seveso, Italy 55
Shanghai 65
Shevardnadze, Edward 127, 158
Sierra Club 19, 57
Silent Spring (Carson) 18–19, 75, 77
Simon, Julian 29n
Single European Act (SEA) 112–13
South Korea 66, 102
South Pacific Forum 167, 180
Soviet Union (USSR) 37, 73–4, 158, 169, 186
  see also Chernobyl, Ukraine, disaster; communist bloc; Russia
Spain
  CO_2 emissions 169, 176
  industrial pollution 65, 66, 111, 113–14
  water 5n
Sri Lanka 86
states, nation 237–9
  authoritarianism 84–7
  interaction 1–2, 238–9, 247, 252–4
  role 7–10
  sovereign rights 45–6
  see also governments
Stockholm Conference *see* United Nations Conference on the Human Environment
Stockholm Declaration 44–6, 49
  principle *21* 45–6, 233
Strong, Maurice
  Stockholm Conference 37, 48
  UNCED 210–11, 220, 221, 231, 233
sulphur dioxide emissions 52, 55, 59–60, 71, 72, 74
  acid rain 104–7, 241

Index

30% Club 12, 106, 111, 253, 257
Sundsvall, Sweden, *1990* meeting at 182–3
Sununu, John 171, 172, 189, 190
Sweden 54
  acid rain concern 34, 41, 104–5
  green party 126
  IPCC meeting in 182–3
  nuclear power 117
  UN conference hosts 105
Switzerland 15

Taiwan 117
taxation 54, 178
technology
  progress 240n
  transfer to developing countries 142–3, 219–20, 228–9
Thailand 64, 67, 102, 151
Thames, river 21
Thatcher, Margaret 129–30, 171, 223
Third World *see* developing countries
30% Club 12, 106, 111, 253, 257
Three Mile Island, Pennsylvania 55, 116
timber 148
  from tropical forests 152, 153–4, 157, 263, 264
Time magazine 19, 127
time scale 253–4
Tolba, Mustafa 141, 142, 184, 205
Toronto Conference on Changing Atmosphere 166
Torrey Canyon 22, 32, 92
tourism 199n, 249
TOVALOP 32, 147
toxic waste disposal 127, 131–3
  Basle Convention 132, 133
  nuclear 3, 21, 115–16, 117–18
  at sea, dumping 21, 32–3, 41–2, 98–9, 118
  treaties 90, 92
Toynbee, Arnold 238
trade, international 2, 8, 263–6
  CFCs 141–2
  ivory 16, 101, 103
  timber 152, 153–4, 157, 263, 264
  toxic waste 131–2
Trail Smelter case 16
treaties and conventions, international 10–12, 15–16, 89–91, 123
  atmospheric pollution 104–7
  endangered species 100–4
  European Community 107–15

marine pollution 91–100
North–South agenda 118–22
nuclear 115–18
Tropical Forestry Action Plan (TFAP) 153–4, 248
Turkey 59, 176

U Thant 27–8, 48
UNESCO 34, 47, 50
United Kingdom (UK)
  acid rain policy 106, 128, 146
  CO_2 emissions 174, 176
  environmental concern 11, 104, 126, 128
  NGOs 257
  nuclear issues 117–18
United Nations (UN)
  agencies 47–9, 50, 217–19, 259–61, 267
  Charter 1
  General Assembly: *1968* 34; *1969* 27–8; *1970* 36; *1987* 128; *1988* 160
  see also names of agencies and conferences
United Nations Commission on Sustainable Development (CSD) 219, 228, 266
United Nations Conference on Desertification (UNCOD) 121, 123
United Nations Conference on Environment and Development (UNCED; Rio de Janeiro, *1992*) 12–13
  effects 266
  gathering 223–30
  lessons 231–5
  planning and preparations 160–2, 207–21
  'Agenda 21' 213–14
  Earth Charter 214–15
  national objectives 210–13
  structure of negotiations 161, 210–13
United Nations Conference on New and Renewable Sources of Energy 121
United Nations Conference on the Human Environment
(Stockholm, *1972*) 12, 13, 51, 63
  Action Plan 44, 46–7, 49
  agreed texts 44–7
  assessment and effects 49–50, 70, 266
  atmospheric pollution 104–5
  compared with UNCED 232–4
  Declaration 44–6, 49
    principle *21* 45–6, 233
  form 42–4

institutional outcome 47–9
North–South divergence 36–41, 120
nuclear power 118
preparation 34, 35–42
resolutions 44, 47–9
10th anniversary meeting, *1982* 121–2
wildlife conservation 100–1
United Nations Conference on the Law of the Sea (UNCLOS) 119, 123
United Nations Environment Programme (UNEP) 232–3, 259
biodiversity 162, 200, 208
CFCs 137, 145
climate change 165, 185
developing countries' attitude 185
Economic Panel Report 141
effects 266
Environment Fund 48, 208
established 47–50
Regional Seas Programme 96–100
sulphur dioxide 71, 217
10th anniversary 128
toxic waste disposal 132
United Nations Fund for Population Activities (UNFPA) 63, 64
United Nations Security Council 267
United States of America (US)
agriculture 79–80
aid donations 226–7
CFCs 105–6, 135–6, 140, 142–3, 144
climate change policy 166–7, 168, 170–3, 184, 185–6, 189, 190, 193, 194
conservation policy 204–5
environmental concern 16, 23, 26–7, 53–4, 57, 125, 171, 243, 255, 256, 257
spending 53, 170, 172
international agreements 260n
legislation 30
Montreal Protocol 142–3
NGOs 256, 257
oil spills 94
Senate, Hansen testimony 166, 170, 255
and Soviet Union 158

UNCED 191, 205, 220, 221, 224, 226–8, 234–5
world influence 69
USSR *see* Soviet Union

Vienna Convention on Protection of the Ozone Layer 137–8
Vietnam War 25, 42
Villach workshop 165
vinyl chloride 57

Wall Street Journal 5
war, environmental effects 212
Ward, Barbara 36
Washington conference, *1973* 101
waste disposal *see* toxic waste disposal
water
drinking 243, 256
see also marine pollution; rivers; seas and oceans
whales and whaling 3, 16, 42, 43, 212, 243
wildlife conservation *see* biodiversity; conservation
World Bank 217, 261–2
Brazil loan 155, 156
development report 52
Stockholm conference 47
UNCED 230, 232n
see also Global Environment Facility
World Charter for Nature 122
World Climate Conferences
First, *1979* 165
Second, *1990* 173, 176, 183–5, 186
World Commission on Environment and Development 122, 128
World Conservation Movement 199
World Conservation Strategy 122, 200, 248
World Health Organization (WHO) 47, 59, 60
World Meteorological Organization (WMO) 165
World Wide Fund for Nature 19
World Wildlife Fund 100, 126
Wuhan, riot 65

Zimbabwe 103, 128